Russia's Military Revival

Russia's Military Revival

Bettina Renz

polity

First published in 2018 by Polity Press

Polity Press
65 Bridge Street
Cambridge CB2 1UR, UK

Polity Press
101 Station Landing
Suite 300
Medford, MA 02155, USA

ISBN-13: 978-1-5095-1614-8
ISBN-13: 978-1-5095-1615-5(pb)

A catalogue record for this book is available from the British Library.

Typeset in 10 on 16.5 Utopia Std by
Servis Filmsetting Ltd, Stockport, Cheshire
Printed and bound in Great Britain by CPI Group (UK) Ltd, Croydon

The publisher has used its best endeavours to ensure that the URLs for external websites referred to in this book are correct and active at the time of going to press. However, the publisher has no responsibility for the websites and can make no guarantee that a site will remain live or that the content is or will remain appropriate.

Every effort has been made to trace all copyright holders, but if any have been inadvertently overlooked the publisher will be pleased to include any necessary credits in any subsequent reprint or edition.

For further information on Polity, visit our website:
politybooks.com

Contents

Acknowledgements

Part of the research and writing for this book was completed during a project funded by the Finnish Prime Minister's Office, Government's Analysis, Assessments and Research Activities fund from October 2015 until July 2016. I would like to thank the fund for the generous support. The project, entitled 'Russian hybrid warfare', was conducted jointly with Hanna Smith at the University of Helsinki's Aleksanteri Institute. Hanna also contributed with her expertise on Russian foreign policy and history to the first chapter of this book. I would like to thank Hanna for her invaluable input, our epic discussions and for her friendship throughout the years. For six months of the project I was based at the Aleksanteri Institute as a senior researcher and I benefited greatly from the positive atmosphere and from the space to think and write that the institute offers to its scholars. I am grateful to everybody there for their ongoing support and friendship. I would like to thank the project's panel of experts for their vital input and for the time they spent with us in Helsinki: Tor Bukkvoll, Samuel Charap, Antulio J. Echevarria II, Keir Giles, Sibylle Scheipers, Hew Strachan and Rod Thornton. Thanks also go to Mikko Lappalainen for his involvement and support.

I am extremely grateful to Louise Knight and Nekane Tanaka Galdos at Polity for their wonderful support, advice and patience throughout the process of writing this book. Both went above and beyond to ensure that the project would come to fruition. I could not have wished for better editors and the book would not have been possible without you! I would like to thank Edwin Bacon, Lance Davies, Matthew Rendall, Rod Thornton, Hanna Smith, Jeremy Smith and Aaron Bateman for reading and offering valuable comments on various parts and chapters of the book. I am also grateful for the constructive criticism and helpful suggestions by the anonymous readers of the manuscript.

Thanks are due to many other people, who are too numerous to list here. I greatly appreciate the time given by the interviewees in Moscow to meet and discuss the subject in spring 2016 and to the many Russian scholars, analysts, journalists, politicians and officials that agreed to speak to me over the years. They all, without exception, have shaped and informed my understanding of Russian politics and military affairs. I would also like to thank all my wonderful colleagues and friends at the University of Nottingham's School of Politics & International Relations. Their support has been vital, especially during the final months of completing the book. I continue to be grateful to everybody, past and present, at the University of Birmingham's Centre for Russian and East European Studies, my intellectual home. Long may it continue! Special thanks go to Edwin Bacon, Julian Cooper, Sarah Whitmore, Mary Buckley, Alex Danchev and Vivien Lowndes, all of whom have been hugely supportive of me and my work. I would not have got here without you.

Finally, my thanks and love go to Jason Curteis and to my parents, Karl-Dieter and Marlene Renz. My parents have always believed in what I am doing and given me tireless encouragement and support, so I would like to dedicate this book to them.

Introduction

"Mr President, acting on your decision, since the 30th, we have been carrying out missions to strike ISIS, Jabhat al-Nusra, and other terrorist groups present on Syrian territory. Since September 30, we have conducted strikes against 112 targets. We are increasing our strikes' intensiveness. Our various intelligence and reconnaissance forces have been working intensively over these last two days and have identified a large number of ISIS targets: command posts, ammunition depots, military hardware, and training camps for their fighters. Vessels from our Caspian Fleet joined our aviation in attacking these targets this morning. Four warships launched 26 Kalibr cruise missiles against 11 targets. Our target monitoring data shows that all targets were destroyed and civilian facilities were not damaged in the strikes. These strikes' results demonstrate the high effectiveness of our missiles launched from a big distance of nearly 1,500 kilometres. This morning, 23 attack aircraft also continued their strikes against insurgent positions. Since September 30, we have destroyed 19 command posts, 12 ammunition depots, 71 pieces of military hardware, and six explosives production workshops producing explosives for car bombs and so on. We are continuing our operations according to plan."

These are the words of the Russian Defence Minister, Sergei Shoigu, reporting to the Commander-in-Chief of the Russian Armed Forces, President Vladimir Putin, exactly one week after Russia's air campaign over Syria was launched on 30 September 2015 (Shoigu 2015). Only ten years prior to this report, such an account of Russian military activities would have appeared like nothing but fiction. Throughout much of the 1990s and 2000s, the Russian armed forces had been left to fall into a state of serious disrepair. As Russia entered the new millennium it appeared clear, as Eugene Rumer and Celeste Wallander wrote that it did so with 'its capacity to project power beyond its borders vastly reduced and its ability to defend its territorial integrity and sovereignty severely tested' (2003: 61). By the middle of the 2000s, many believed, both in Russia and in the West, that the ongoing neglect of the Russian armed forces had pushed them close to irreversible ruin. Given that their service personnel were by now 'impoverished, demoralized and largely ineffective' (Barany 2005: 33) and the forces 'woefully inadequate to address the country's security threats' (Golts and Putnam 2004: 121), it seemed clear that Russia no longer cast the shadow of a global military power.

Against this background Russia has experienced a remarkable military revival within barely more than a decade. The operations in Syria demonstrated that many of the shortcomings, which had led to humiliating defeat in the first Chechen War and to operational problems in other conflicts, had been decisively overcome. One year into the Syria intervention in autumn 2016, Russian forces had experienced minimal losses, both in the air and on the ground. Moscow had put on display new capabilities, such as vastly improved command and control and inter-service coordination, as well as advanced technologies like precision-guided munitions, including cruise missiles fired from the Caspian and Mediterranean seas. What came, perhaps, as the biggest surprise to many observers was that Russia now had the sealift and airlift capabilities required to launch military operations far beyond its immediate neighbourhood (Gorenburg 2016). As such, in Ruslan Pukhov's words, Russia's air operations over Syria represented 'the most spectacular military-political event of our time' (2016).

The world's amazement at the Kremlin's conspicuous display of its shiny new military power in Syria did not come completely out of the blue. This operation was launched only a year and a half after another Russian surprise military success: the annexation of Crimea in spring 2014. In this operation, Moscow had demonstrated not so much the advances it had made in the procurement and use of modern technology and its ability to launch a twenty-first-century air campaign. The Crimea operation, instead, had stood out for extreme restraint in the application of any physical violence. 'Little, green and polite' special operations soldiers (Nikolsky 2015), in combination with an information campaign and other non-physical tools, allowed Russia to achieve its objectives without almost a single shot being fired. This stood in stark contrast to previous Russian military operations, which all, without exception, had been criticized for the excessive use of force. Until Crimea it had been widely assumed that Russian military strategists were unable to move beyond Cold War thinking on large-scale inter-state warfare. The approach in Crimea, which later became known as 'hybrid warfare', suggested that serious advances had been made also in Moscow's strategic thought. Whilst previous conflicts were approached as conventional warfare campaigns almost irrespective of the circumstances, in Crimea appropriate means were skilfully matched to the conflict's ends. After years of failed attempts at reforming the Russian armed forces, the 2008 modernization programme finally led to systematic change and restored the country's standing as a serious military actor. As *The National Interest* put it, 'Russia's military is back' (Gvosdev 2014).

Reactions to Russia's military revival

For much of the post-Soviet era, it appeared clear that Russia's days as a serious global military player were over for good. Even throughout the troubled 1990s, Moscow had maintained one of the world's largest nuclear arsenals. This continued to give it some of the prestige and privileges afforded to military great powers, such as a permanent seat on the UN Security Council. However, in an age of small wars and insurgencies, where state-on-state warfare appeared to be a thing of the past, a strong nuclear deterrent alone was increasingly seen to be of little more than symbolic value. Within the former Soviet region, Russia always remained by far the most dominant military actor. It used this strength with impunity there in various conflicts since the early 1990s. Operations beyond this region, however, were largely outside of the realm of its possibilities, and its conventional capabilities were no match for the much more advanced militaries of the West. International views of Russia's military, as a mere shadow of its former Soviet self, changed almost overnight with the annexation of Crimea. Surprise turned into awe as the operations in Syria unfolded. These not only showed that the country's military capabilities had dramatically improved. They also demonstrated that Moscow was now confident and willing to use military force to pursue its interests on a global level, irrespective of strong condemnation by the West. What marks the military revival as a significant turning point in post-Cold War global security is the fact that, for the first time since the collapse of the Soviet Union, a militarily resurgent Russia is seen as a threat not only to its neighbours, but also to the West.

Russia's military aggression against Ukraine in 2014 took most countries by surprise (House of Lords 2015: 6). As Viatcheslav Morozov wrote, it 'created a shockwave in the European security system. It suddenly became apparent that certain key rules of international conduct in Europe could no longer be taken for granted' (2015: 26). There was a sense that the Kremlin's actions were the result of a relatively sudden and dramatic change in foreign policy – a 'paradigm shift' – which, enabled by a revived military, signified a 'seismic change in Russia's role in the world' (Rutland 2014). Having previously paid little attention to the extensive modernization the Russian armed forces had been undergoing for some time, questions started being asked by many observers and officials in the West about the purpose of this undertaking. Many believed that the only explanation for the Kremlin's efforts to strengthen the country's military capabilities was the intention to engage in further aggressive action. As Jonathan Masters wrote for the US think tank, the Council on Foreign Relations, 'the Russian armed forces are in the midst of a historic overhaul with significant consequences for Eurasian politics and security' (2015).

To many observers, Putin's intentions seemed to be crystal clear. Most immediately, the 'paradigm shift' in the Kremlin's ambitions posed a threat to Ukraine, where the annexation of Crimea was only the beginning. In late March 2014, US intelligence officials reportedly cautioned that the probability of a full invasion of Ukraine 'was very high' (Gover 2014). A year after the annexation, some analysts were still convinced that Putin was driven by the desire to gain more territory. As Hans Binnendijk, a senior fellow at the Johns Hopkins Center for Transatlantic Relations, and John Herbst, a former US ambassador to Ukraine, wrote about Putin in the *New York Times*, 'his long term goal may be the creation of "Novorossiya", or New Russia, which would constitute all of southern Ukraine past Odessa to Moldova, and would enable Russia to control the entire northern coast of the Black Sea. There are no large armies to stop him' (Binnendijk and Herbst 2015). Many believed that 'Russia's military buildup is a harbinger of neo-imperial expansion', where the annexation of Crimea was merely a first stroke of the brush on a vast canvas (Ramani 2016). As former US Secretary for Defence Leon Panetta put it, 'Putin's main interest is to try and restore the old Soviet Union. I mean, that's what drives him' (quoted in CSIS 2016).

Fears were also expressed that, what many saw as the Kremlin's new expansionist vision, might extend much further and even NATO territory might not be off limits. The Baltic States were seen to be particularly threatened by this. As a journalist writing for the UK broadsheet the *Daily Telegraph* asserted, if Putin 'concludes that his adventure in Ukraine has served Russia's interests, then he will turn on new targets – and the trio of countries along the Baltic coast would probably be next' (Blair 2015). The US military analyst and Russia expert, Stephen Blank, in 2016 was 'counting down to a Russian invasion of the Baltics'. The British Defence Secretary, Michael Fallon, also believed that there was a 'clear and present danger' that the Baltic States would be Moscow's next target (Farmer 2015).

The Kremlin's aggressive use of its improved armed forces in Crimea was seen by many not only as a danger to Russia's immediate neighbours, but to the whole of Europe, and even to the United States and international security at large. As then-US Defense Secretary Chuck Hagel asserted in November 2014, Russia had been investing in its armed forces 'to blunt our military's technological edge . . . If this capability is eroded or lost, we will see a world far more dangerous and unstable, far more threatening to America and our citizens here at home than we have seen since World War II' (2014a). Gustav Gressel, writing for the European Council on Foreign Relations, noted that Europe was in need of finding a response to 'Russian expansionism'. Although he conceded that 'a major escalation' on the European continent was 'not imminent', urgent action was required, because 'Russia is clearly preparing itself for offensive operations' (2015: 1, 13). In the words of Damon Wilson, a former national security aide to President George W. Bush, 'Putin just declared war on the European order and it's demanding that the United States focus on Europe again as a security issue' (quoted in Shear and Baker 2014).

When Russia launched its first airstrikes in Syria in autumn 2015, this affirmed in the eyes of many observers that a militarily resurgent Russia posed a security threat of global dimensions, as Putin continued his 'power play', this time in the Mediterranean (K. Johnson 2015). As the former secretary general of NATO, Anders Fogh Rasmussen, wrote in spring 2016, 'we should hold no illusions about Moscow's intentions . . . The clash is not only taking place in our shared neighbourhood. Moscow clearly aims to undermine the liberal international order and Western unity that has served us well since the end of World War II'. An article in *The National Interest* asserted that 'Russia's ongoing military buildup in Syria poses a serious challenge to American policy in the region' (Graham 2015), others interpreted Moscow's 'unanticipated military foray into Syria' as a 'proxy US-Russian conflict' (Stent 2016: 106). A journalist writing for the *New York Post* believed that Putin was intent not only on supporting the regime of Syrian President Bashar Assad, but on controlling Syria as a whole: 'The Syrian coast will become another Crimea, if not completely annexed, at least occupied . . . Putin has arranged it so that no matter what happens in Syria, he wins – and we lose' (Taheri 2015). In a strategy paper issued by the United States European Command in October 2015, a 'revanchist Russia' was listed as the top threat to European security, to the US homeland and to global stability (Breedlove 2015: 1). In 2016, Polish Prime Minister Antoni Macierewicz called Russia 'the biggest threat to global security today' (Sharkov 2016).

The demonstration of Russia's new military prowess in both Ukraine and in Syria has led to fears that the previously superior militaries of NATO and of the US were in real danger of being over-taken, with severe repercussions for security in Europe and beyond. As Karl Nerenieks, a retired Major General of the Swedish Armed Forces, remarked in 2014, Russia's armed forces 'regained their capability to mount large conventional military operations. They are, I would say, some years ahead of us if we started to train for the same thing today' (House of Commons Defence Committee 2014a). As Blank asserted, Russia's military revival was being watched closely by the US Strategic Command, because of the belief that 'within five years Russia could run multiple Ukraine-sized operations in Europe'. Blank (2015) also noted that, if Russian procurement plans were carried through to 2025, 'this force would have parity with the US and NATO in conventional and nuclear dimensions of high-tech warfare'.

The way in which Crimea had been taken raised concerns that innovations in Russian strategic thought meant that the country had developed a new 'hybrid' approach to warfare, which the West already was unable to stand up against. To quote Chuck Hagel again, Moscow was developing 'capabilities that appear designed to coun-ter traditional US military advantages' (2014b). This was also noted by Blank (2015): Moscow 'now seems to favour an approach based on hybrid or multidimensional warfare, similar to the Chinese concept of "unrestricted war", embracing simultaneous employment of multiple instruments of war, including non-military means where information warfare, such as mass political manipulation, is a major capability'. Michael Gordon, writing about the Crimea operation for the *New York Times*, described a military 'skilfully employing 21st century tactics that combine cyberwarfare, an energetic information campaign and the use of highly trained special operation troops to seize the initiative from the West'. This, he believed, had 'implications for the security of Moldova, Georgia, Central Asian nations and even the Central Europe nations that are members of NATO' (Gordon 2014). As a UK House of Commons Defence Committee report asserted, the 'new and less conventional military techniques' Russia had developed 'represent the most immediate threat to its NATO neighbours and other NATO Member States' (House of Commons Defence Committee 2014b).

Since the events in spring 2014, it has been widely accepted that improvements to the West's defence posture and planning have become vital. In particular, European states have to increase their defence spending, which had been affected by significant cuts since the end of the Cold War. As the *New York Times* noted, this had rendered 'NATO less formidable as deterrent to Russia' (Cooper and Erlanger 2014). As then-NATO Secretary General Rasmussen stated in March 2014, 'developments in Ukraine are a stark reminder that security in Europe cannot be taken for granted . . . That is why I will continue to remind European nations that they need to step up politically and militarily. To hold the line on defence cuts. To increase their defence spending'. At a NATO summit held in early autumn in 2014, alliance members pledged to freeze cuts and, where necessary, to increase spending to two per cent of their respective GDPs within the next decade. By 2016 the implementation of this pledge was well under way (Jones 2016). European states outside of the NATO alliance, including Finland and Sweden, have also increased their military budgets in view of the developments in Moscow (Tiessalo 2015; Rathi 2016).

Words of warning about the dangers of a militarily resurgent Russia have resulted in various actions intended to secure Europe. NATO has stepped up its force posture in Central and Eastern Europe, and especially in the Baltic States. The alliance has also held a number of military exercises to demonstrate resolve and unity, and to reassure those member states in the region that are particularly worried about the potential for aggressive Russian action. Whilst remaining outside of NATO, Finland and Sweden have stepped up cooperation with the alliance and both have sought defence cooperation agreements with the US (Borger 2016). Moreover, Nordic countries, including Norway, Denmark and Iceland, have increased their own military cooperation to enhance preparedness for any hostile Russian act in the Baltic region (Agence France Press 2015). There is a feeling in the West that the Kremlin's actions since 2014 have left no option but to take an uncompromisingly tough stance and to bolster military postures and defences in order to deter an aggressive Russia. As the former chief of staff of the British armed forces, Richard Dannat, noted, 'with a resurgent Russia this is a poor moment for the US-led West to be weak in resolve and muscle'. Sanctions and diplomacy were insufficient as a response to Moscow's actions, in his eyes, because Putin 'will look beyond those things to see where the real check on his actions might come from' (quoted in Cooper and Erlanger 2014). This was echoed by NATO Supreme Allied Commander Europe (SACEUR) General Curtis Scaparrotti, who put forward an unambiguous warning to the alliance in May 2016: 'a resurgent Russia [is] striving to project itself as a world power . . . To address these challenges, we must continue to maintain and enhance our levels of readiness and our agility in the spirit of being able to fight tonight if deterrence fails' (quoted in Bodner 2016).

The purpose of this book

This book argues that reactions of this nature to the annexation of Crimea and to Moscow's involvement in the war in Syria are problematic, because they are based on three misguided assumptions regarding the timing, purpose and scope of Russia's military revival: first, the view that the desire for a powerful military and its use signals a 'paradigm shift' in the Kremlin's outlook; second, the idea that the reason the military revival is pursued necessarily is to enable an expansionist and aggressive foreign policy; and third, the notion that Russian military capabilities now rival those of the West. These assumptions are misguided, because they fail to take into account the historical and international context of the military revival, which did not occur in a vacuum. The purpose of the book is to provide this context.

Taking these assumptions as a starting point, the book's analysis revolves around three major arguments. The first argument purports that the Kremlin's most recent efforts to strengthen the armed forces is not the result of a 'paradigm shift' in views on the utility of military power. This is because the decision to modernize the armed forces did not come out of nowhere or occur in a vacuum. As the chapters of the book will show, rather than representing a break from the past, recent developments are the result of a complex confluence of historical, political and economic factors, many of which have been long in the making. Second, the modernization of Russia's military in recent years was not determined by the desire to pursue expansionist or unilaterally aggressive policies in a bid for domination. The assumption that this is the case reflects a one-sided understanding of why states, including Russia, view a powerful military as an important asset. Military power is a flexible tool of statecraft and its utility is not limited to the fighting of wars and defeating of opponents. This needs to be borne in mind when the Kremlin's reasons for strengthening its armed forces are assessed. As such, recent efforts to reform the country's armed forces can only be understood within the context of the variety of functions the Russian armed forces have fulfilled throughout history. Third and finally, the book argues that the military revival has not resulted in capabilities that have substantially altered the power balance in Europe or even beyond. This is because a state's military power is never absolute, but always relative to that of others. It is beyond doubt that Russia's military capabilities today are much improved compared to what they were in the 1990s. In terms of military planning, too, there have been marked improvements in the ability of Russian strategists to fine-tune military tactics to suit the circumstances of operations of various intensity. However, these improvements do not mean that the country's capabilities now rival those of the West, will guarantee victory in all cases, or even that they have created substantially new opportunities for the achievement of objectives that were not achievable before.

Improvements in Russia's military capabilities and Moscow's growing confidence in using armed force as an instrument of foreign policy are significant and this poses challenges to its neighbours and to the West. However, the precise nature of these challenges is not as straightforward as often implied. The book's arguments are developed in five chapters in order to provide detailed context for an informed assessment of recent events. The topics of these chapters are the role of the military in Russian foreign policy in the past and today, reforms of the Russian armed forces since the early 1990s, the significance of the force structures as an important component of the country's military establishment, Moscow's uses of military force in wars and conflicts since the end of the Cold War, and developments in the country's military thinking.

A contextualized analysis of Moscow's reasons for strengthening its armed forces, and of the significance of this for the security of both its neighbours and of the West, is not only of interest as an exercise of academic inquiry. It also has substantial policy relevance. Following the annexation of Crimea, it has become a widely accepted fact that a lack of capacity to understand political developments in contemporary Russia caused the West to 'sleep walk' into the current crisis (House of Lords 2015: 6; Monaghan 2016: 26–7). A contextualized study of Russia's military revival contributes to a better understanding of the Kremlin's thinking and actions, which can help to ensure that any potential future actions will come as less of a surprise. A better understanding of the reasons for, and implications of Russia's military revival is also significant for policies adopted by the West vis-à-vis a more assertive Kremlin. If such policies are insufficiently informed by an awareness of the motivations driving Russian behaviour, they could fail and inadvertently lead to spiralling tensions. A contextualized understanding of the military revival is essential not to justify Moscow's actions, but rather it is vital to inform policy and decision makers in the West where to go from here. As Carolina Vendil Pallin noted, 'what the West does will matter . . . In spite of the fact that change must come from Russia, the policy response of the EU and

NATO matters – and it does so irrespectively of how events in Russia develop' (2015: 14, 20).

The book's five chapters each address in detail an aspect of military power that is important for an informed understanding of Russia's military revival. The chapters have been written so they can also be read individually. Read as a whole, they provide a comprehensive context for a better understanding of the timing, intentions and scope of Moscow's efforts to restore its country's armed forces and of the implications this has for international security.

Chapter 1 outlines the role of military power in Russian foreign policy in the past and today. As such, the chapter delivers the historical background of the book. The chapter is structured around four factors, all of which have been important in shaping foreign policy since the time of the Russian Empire. These are great power status, sovereignty, imperialism/imperial legacy, and multilateralism. Although these factors are not the only 'persistent factors' in Russian foreign policy, they are particularly salient for explaining the role of military power within it. The chapter demonstrates that not only has a strong military always been an essential component in Russia's self-perception as a great power, but also that Moscow has always viewed and used the armed forces as a flexible tool of foreign policy. Highlighting relevant patterns and developments in the past, the chapter suggests that preparation for offensive war and expansion is unlikely to be the most important reason for the recent military revival.

Chapter 2 traces Russia's struggle to transform what was left of the former Soviet military into a force fit for the twenty-first century. It shows that the neglect of the armed forces during the Yeltsin years was the result of a complex combination of political, societal and financial factors and not the result of a principled decision. Central elements of the 2008 modernization programme were debated since the early 1990s and Moscow never abandoned its ambition to be a global military power. Aided by a much improved economic situation, the Russian military's fortunes started to turn when Putin became president and made the military reform agenda a priority. The chapter details the impressive advances in Russian military capabilities that have been made since 2008 and were demonstrated on a limited scale in Crimea and in Syria. It also shows, however, that the modernization process is far from complete and Russia is still a long way off meeting its goal of parity with other leading powers. Important obstacles continue to stand in the way. These include ongoing problems with maintaining the desired level of manpower and the defence industry's inability to deliver the quality of technology required for competing with the world's most advanced nations. Russia's deepening economic crisis since 2009 has also meant that the affordability of its ambitions is far from certain.

Chapter 3 provides an overview of Russia's force structures. These include, amongst others, the Interior Ministry (MVD), the Ministry for Emergency Situations (MChS), the Federal Security Service (FSB) and the National Guard Service (FSNG). The force structures form an important component of the country's military establishment, but are often ignored in analyses, because they do not fit easily into Western frameworks. The chapter argues that the fate of the force structures since 1991 offers important insight into the timing and reasons for the revival of Russian military power in recent years. The creation of the force structures was determined in part by the need to build capabilities for dealing with a range of new security challenges that the regular armed forces were ill-equipped to deal with. Force structure personnel were used in various low-intensity missions since the 1990s and have also participated in multilateral security cooperation, including with NATO, throughout the 2000s. Political motivations, however, have been the most important reason for maintaining these structures, which are tasked predominantly with internal security. The availability of force structures loyal and subordinated directly to the president was as important for Yeltsin as it is for Putin today. They are essential for ensuring internal order and regime stability, which are increasingly viewed as being under threat and thus as a matter of national security.

Chapter 4 discusses the annexation of Crimea and the intervention in Syria within the context of the Kremlin's other uses of military power since the early 1990s. It argues that there is little evidence to suggest a fundamental change in Moscow's views on the utility of force, or that the desire to expand its territory or to confront the West in a bid for domination have become the major drivers. Since the early 1990s, Russia has been using military force in pursuit of a variety of policy objectives. With regard to its neighbourhood, the imperial legacy has informed its decision to use force in the region since the end of the Cold War. However, chance and contingency, status concerns, insecurity and strategic interests have also been important. If patterns from the past are anything to go by, further expansion is fairly unlikely, because more indirect forms of domination offer a lever of control that is more valuable than adding more territory to an already vast state. With regard to the West, Russia has become more assertive in pursuing an independent foreign policy, even if this risks a breakdown in relations. That said, Russia's approach to the West continues to be characterized by a complex interplay of cooperation and conflict. It desires inclusion, but does not avoid confrontation when it feels that its views are not taken into account.

Chapter 5 assesses developments in Russian military thinking since the end of the Cold War with a particular emphasis on the 'hybrid warfare' discussion. The chapter argues that the 'hybrid warfare' debate does not adequately reflect developments in Russian strategic thought and misrepresents the country's ambitions as a global military actor. The perception of Russian backwardness in military thinking during the 1990s and 2000s was exacerbated by the preoccupation in Western strategic thought with counterinsurgency warfare, which pushed the perceived relevance of 'traditional' war-fighting onto the backburner. In Russia, the idea that conventional warfare was a thing of the past never established itself as a consensus view, because it did not correspond to the country's strategic priorities. The chapter also engages with the problems pertaining to the concept of 'hybrid warfare' as an analytical tool. Russia's approach in Crimea showed that the country had vastly improved its ability to fine-tune military tactics to the requirements of different conflict scenarios. However, it has not found a new key to military success in the form of 'hybrid warfare'.

The conclusion summarizes the book's main arguments. Returning to the three assumptions in Western reactions outlined above, it suggests that a more informed understanding of the timing, reasons and scope of the military revival puts into question the sometimes alarmist interpretations of Russia as an imminent threat to international security. The Kremlin has become more assertive and better military capabilities have offered it more opportunity to use force in the future. This does not mean, however, that stronger armed forces automatically signal Putin's desire to pursue expansionist policies. Moscow has always viewed a strong military as essential, so the revival was only a matter of time. Russian foreign policy is determined by drivers that are much more complex than the simple wish for global domination. In any case, the country's relative military power is still limited in many respects. The conclusion also considers possible options for neighbouring states and for the West in responding to a more assertive and militarily capable Russia. There is no easy way out of the significant breakdown in relations since 2014. However, it is clear that policies based on a lack of understanding of the Kremlin's intentions will significantly increase the danger of tensions spiralling out of control.

Chapter 1

Russian foreign policy and military power

With Hanna Smith

This chapter outlines the meaning of military power in Russian foreign policy throughout its history. An understanding of this is important for a nuanced assessment of recent developments and of the potential future implications of stronger Russian armed forces. The process of military modernization since 2008 did not occur in a vacuum and can only be understood within the context of relevant historical developments, political processes and foreign policy priorities. The nexus between military power and foreign policy is obviously important. However, to date this link has rarely been addressed explicitly or systematically in studies of the military in post-Soviet Russia. Major studies of the subject have tended to focus on capabilities, leadership, civil-military relations or organizational changes, but they have rarely systematically studied the role of the military as a tool of Russian foreign policy (see, e.g., Arbatov 1998; Aldis and McDermott 2004; Golts and Putnam 2004; Barany 2005; Vendil Pallin 2009). Whilst the study of capabilities and effectiveness of the Russian military is essential, it tells us little about Moscow's intentions to use these capabilities. It is the Kremlin's possible intentions that have led to fears about the country's resurgence since 2014.

Estimating a country's intentions regarding future uses of military force is not straightforward. As John Mearsheimer put it, 'unlike military capabilities, intentions cannot be empirically verified. Intentions are in the minds of decision makers and they are especially difficult to discern' (2010: 79). Although the Kremlin's intentions cannot be established for certain, studying the significance of military power in Russian foreign policy throughout history can flag up certain patterns and developments that give an indication of possible future trends, likelihoods and possibilities. The chapter suggests that Russia has always viewed a strong military as important. However, the reason for this was never limited to the desire to pursue aggressive or expansionist policies. As Robert Art noted, military power is a pliable tool of statecraft that cannot be divorced from foreign policy. As such, the utility of military power is not limited to the fighting of wars and defeating of opponents. In Art's words, 'force is "fungible". It can be used for a wide variety of tasks and across different policy domains; it can be employed for both military and non-military purposes' (1996: 8). This is important also for the study of military power in Russian foreign policy. As the chapter will show, today, as it has done in the past, the Kremlin views this power as a flexible tool that can be used in symbolic and physical ways in pursuit of specific objectives. This puts into question the interpretation of the recent military revival as a necessary indication of the Kremlin's intent to expand or to launch a bid for global domination.

The chapter is structured around four broad issues that are widely recognized as important for a nuanced understanding of Russian foreign policy. The issues addressed in this chapter are not the only factors that have shaped Russia's international politics throughout its history. Alfred J. Rieber has previously identified what he called 'persistent factors' as a starting point for analysing Russian foreign policy trends: relative economic backwardness, porous frontiers, a multinational society and cultural alienation (1993). This chapter shares the view that there are persistent factors in Russian foreign policy, which, as Rieber has put it, 'are the result of an evolutionary process, features of historical life that have their origins in the period of state formation in the fifteenth and sixteenth centuries' (2007: 206). For the purpose of this chapter, however, the focus is on an adapted set of factors that addresses those of Rieber indirectly, but is more suitable for the study of the military in Russian foreign policy specifically. The factors addressed in the chapter are as follows: the first is Russia's great power status and its quest for international recognition as such a power; second, sovereignty as a factor in Russian foreign policy is outlined with a particular emphasis on the country's specific understanding of the concept; third, Russia's imperial legacy, including imperial expansion, is addressed; fourth and finally, the chapter looks at multi-lateralism in Russian foreign policy.

The significance of these four factors in shaping Russian foreign policy has been highlighted over the years by various scholars writing on Russian history and politics. Leading western historians of Russian national and imperial identity, such as Geoffrey Hosking (1998, 2001), Hugh Seton-Watson (1989), Vera Tolz (2001) and Ronald Suny (2001), have each discussed some of these factors, which, in their eyes, have been drivers of Russian foreign policy since the time of Ivan the Terrible, Peter the Great and the Napoleonic wars. Other scholars, focusing on post-imperial Russia, have variously detected the same factors in Soviet foreign policy making (Dallinn 1960; Ulam 1968; Bialer 1981; Nation 1993) and also in post-Soviet foreign policy (Petro and Rubinstein 1996; Donaldson and Nogee 2009; Legvold 2009; Trenin 2011).

The discussion in this chapter of Russian foreign policy throughout its history is by necessity brief and selective. However, it provides essential context and background for the book's central argument: the recent military revival did not occur in a vacuum. It can only be understood within the context of relevant developments in Russian foreign policy and the role of the military within it.

Great power status

There has always been a strong contradiction between the way others see Russia and how Russia perceives itself. This tension has been one of the central drivers of the country's foreign policy. As Iver Neumann has written, 'from early contacts between Muscovy and the Holy Roman Empire through the rapid increase in contact during and following Peter the Great's reign and finally during the Soviet period, Russia has tried to be recognized by the leading European powers as their equal' (2008: 128). Throughout its existence, Russia occupied vast expanses of the Eurasian land mass and it has been recognized in the past as both an Asiatic empire and a European great power. For some forty years during the Cold War it enjoyed global superpower status. The country's history, strongly influenced by geography and geopolitics, has resulted in the understanding that Russia's destiny is to be a great power. The self-perception as a great power has been a constant feature in the country's identity. This did not change with the collapse of the Soviet Union in 1991, when the Russian Federation emerged as a newly independent country characterized by weak statehood, a struggling economy and dishevelled armed forces. As Margot Light noted, in the years following the end of the Cold War, 'Russia was clearly not a superpower; indeed, it was questionable whether it was a Great power. Yet to ordinary people, as well as to politicians, it was unthinkable that Russia could be anything less than this' (2010: 229).

In contrast to Russia's own views on its place in the world, in the West and elsewhere there was not an automatic assumption that it would inherit the Soviet Union's global power status. As a result, during the 1990s, Moscow's ongoing quest to maintain its great power status based on its historical self-perception was largely sidelined in Western debates. If the issue of status was addressed at all, discussions focused on whether the country could still be regarded as a great power, and if so, on what grounds (Adomeit 1995; Neumann 1996; Hedenskog et al. 2005). Although for Russia itself, accepting the loss of great power status was never an option at any point during the post-Soviet years (Neumann 2008: 128–9), Western observers only started to pay attention to the significance of Moscow's self-perception as such a power during the first decade of the 2000s (Clunan 2009; Light 2010; Feklyunina 2012; Forsberg et al. 2014; H. Smith 2016).

In order to understand the centrality of great power status in foreign policy, an awareness of how the concept is understood in Russia itself is important. *Derzhavnost'* literally translates as 'greatpowerness' and originates in the word *derzhava*. This can signify both 'state' and 'power', so another good translation would be 'the quality of being a power' (Papkova 2008: 70). E. Wayne Merry has described *derzhavnost'* as 'a belief in the primacy and greatness of the Russian state raised almost to the level of a secular religion' (2016: 29). The term has a strong association with the tsarist past and with a powerful state led by a strong ruler. Thus the concept links the past to the present and symbolizes a strong state abroad as well as at home. As Igor Orlov has explained, *'derzhavnost'* should be understood as the characteristics of a country with political, economic, military and spiritual power in the world, as well as the ability to influence and apply pressure in international relations. It is also a specific ideological construct that reflects the extent of consciousness, political weight, economic and military power of a country. It is present in all aspects of society' (2006).

In the post-Soviet period it was General Alexander Lebed who explicitly introduced the concept of great power status into popular political discourse, during his campaign for the presidency in 1996. The concept grounded his electoral campaign in a positive image of the past and provided a future vision for Russia. Great power status, according to Lebed's vision, signified the restoration of Russia's prestige through the creation of conditions in the country that were worth defending and where the state would serve the nation. A strong and effective military, whose major purpose would be to defend the motherland, was a central condition for the achievement of his vision (Allensworth 1998: 51–2, 55). Lebed's popularity, which led some observers to believe that he was the 'warrior who would rule Russia' (Lambeth 1996), demonstrated that his vision resonated positively in Russian society. When Putin was elected president, he too, identified great power status as a central Russian value (Kolstø 2004).

Within this context, Russia's recent military revival needs to be understood as a part of the drive to reassert the country's great power status. Military power is not the only factor on which a country's status in the international system is based, and other strengths, such as economic might and prosperity, are also important. However, military power has always been an indispensable characteristic and symbol for any global power. For Russia, too, having a strong military has been an essential ingredient in the country's quest for international status recognition, especially during periods when it could not compete with other global powers in other areas. It was military might that made the Tsarist Empire and during the Cold War, too, it was the Soviet Union's ability to project military power on a global level that elevated the country to the position of one of the world's two superpowers. During the final years of the Soviet Union, Mikhail Gorbachev sought to steer the country away from its focus on military strength for global competition and set out to pursue a wider variety of tools for reasserting the Soviet Union's status. As a part of his New Thinking, foreign policy concepts like the idea of the 'Common European Home', interdependence, universal values and all-human interests were used to emphasize the country's international status. Gorbachev's understanding that the prohibitive costs resulting from the maintenance of Soviet military power would undermine the country's viability and status in the long term was an important factor in this decision (Brown 1996: 222).

In the early 1990s Yeltsin continued on the course of restoring Russia's great power status not primarily through military might, but through pursuing political stability and cooperation with the West. He, too, was unable to achieve this. Political instability, such as the constitutional crisis in 1993, corruption, military failures in the first Chechen War and an obviously failing economy all projected an image of Russia as a weak state and a country in decline. Both Gorbachev and Yeltsin are still remembered in Russia today for allowing the country's status to decline (H. Smith 2014).

As Russia's conventional military disintegrated throughout the 1990s owing to the lack of funds and systematic reforms, so did the country's international image as a global power. This did not escape the country's political elite, which promised to restore the Russian military to its former glory on many occasions. As explained in the next chapter, however, political leaders were unable to turn this ambition into reality until the turn of the millennium. That said, as discussed in detail in chapter 4, Russia used military power as a tool of foreign policy well before the 2008 modernization programme got under way. It did so in different parts of the Commonwealth of Independent States (CIS) region, in multilateral peacekeeping operations in the Balkans, and, to a limited extent, during the tensions with NATO over the Kosovo War.

When Putin rose to political prominence at the turn of the millennium, he immediately pledged to restore the country's rightful place in the world. In a televised address broadcast on the eve of the presidential elections in March 2000, Putin confirmed that, as future president, his aim was to restore Russia's international standing and the military would play an important role in this process:

> On 26th March we are electing not only the head of state but also appointing the Supreme Commander because the President, by virtue of his office, is simultaneously the Supreme Commander of the Armed Forces. Russia is one of the biggest countries in the world and a strong nuclear power. This is something that not only our friends remember. Let me repeat that we are electing the President, whose duty is to ensure economic recovery, restore the country's prestige and leading role in the world, make Russia governable again, and deliver stability and prosperity to everyone. (Putin 2000)

As the next chapter shows, the Russian military's fortunes changed with the election of Putin to the presidency. This was helped by the rapidly improving economic situation and because the new president made military-related matters a real priority from the outset. Putin's vision of Russia as a great power, which he shared with most of his predecessors, strengthened his resolve not only to declare this a priority, but to ensure that this priority would be met.

In spite of the strong emphasis by Putin on the importance of military power, Western analyses of Russian foreign policy during the first decade of his rule focused on the Kremlin's use of energy policies to seek influence around the world. As it happened, the country's emergence as an 'energy superpower' never materialized as some had expected. The use of gas and oil for the achievement of foreign policy objectives did little to restore Russia's international status and was even counterproductive. As Peter Rutland put it, 'rather than boost Russia's prestige and authority, it has stoked anxiety and driven countries to seek alliances and take other steps to protect themselves from Russian pressure' (2008). Although military modernization had proceeded with impressive speed since 2008, it was not until the annexation of Crimea in spring 2014 that the West's attention turned to the significance of military power in the Kremlin's quest for great power recognition.

Russia's efforts to strengthen its military power in pursuit of international status recognition, today and in the past, have not only been a factor in its foreign policy; they have also had significant domestic dimensions and consequences. Military reforms have had most impact and raised Russia's global status most effectively when accompanied by changes to the internal fabric of the Russian state. For example, Peter the Great's modernization drive at the start of the eighteenth century focused first of all on the army and navy, because a modernized military was seen as an important prerequisite for Russia's status as a great European power. Like many later Russian leaders, Peter harnessed autocratic power to mobilize society and the economy for war (Fuller 1992: 83–4; Snyder 1994: 182). However, his reforms also led to changes in Russian society in the spheres of education, culture, administration, science and technology. In contrast, in the first half of the nineteenth century, Tsar Nicholas I sought to maintain Russia's great power status whilst pursuing conservative domestic policies and resisting any attempts to modernize Russian society and internal affairs. Russia's defeat at the hands of Great Britain, France and Turkey in the Crimean War at the end of his reign showed how the country's comparative power had declined. Internal stagnation had contributed to external problems, brought into sharp focus by a major military defeat. Alexander II, who ruled the Empire from 1855 to 1881, focused first and foremost on domestic issues. But at the same time, Russia exerted itself in efforts to reverse some of the conditions it had been forced to accept following defeat in the Crimean War, including the imposition of a demilitarized zone in the Black Sea area. He also fought a short but successful war against Turkey in the Balkans in 1877–8 (Jelavich 1991: 143–5). Meanwhile, the idea of Russia's destiny as a great power was still strongly expressed in intellectual and political circles.

When he became heir to the throne in 1881, Alexander III sought to revive Russia's status through vigorous efforts to boost his Empire's economic and military power. Although he was viewed as a conservative, Alexander III focused on economic modernization first of all as a way of building Russia's military power and international status. During his reign the system of domestic surveillance grew to new heights, but in foreign policy Alexander sought to avoid conflict, being aware of the country's comparative military weakness. Tsar Nicholas II showed little inclination to modernize the Russian army, but drew heavily on military symbolism and regarded the army as a major bearer of Russian power. This encouraged him to try and boost Russia's status in international affairs, first through war with Japan in 1904–5, which ended in humiliation and was one of the causes of the 1905 revolution in Russia; and second by going to war with Austria-Hungary in 1914 – the First World War – with disastrous results (Jelavich 1974: 189–280).

In the early years of the Soviet Union, the young state faced a struggle for its very survival, preventing it from playing a part in the international great power game. At the same time, its leaders saw themselves as a new kind of vanguard that would lead a world socialist revolution. Both Lenin and Stalin were acutely aware that the country was well behind the Western powers in terms of economic development and military capability. The failed war with Poland in 1920 underlined the fledgling state's military weakness. The Soviet Union needed to modernize and build up its military strength in order to be taken seriously in international politics. Rapid industrialization during the 1930s and, above all, the Red Army's victories over Hitler in the Second World War, meant that the Soviet Union could once again claim to be a great power, and also that its status was internationally recognized. In the Soviet Union, economic and societal transformation did not follow a Western model of modernization, but took a unique path. This reinforced the idea of the Soviet Union as a vanguard state and as a power looked up to by a significant portion of the globe. Ultimately, increasingly fierce military competition with the United States during the Cold War, while the economy was beginning to stagnate, meant that the defence industry and military spending had to be prioritized over all other areas of public spending and investment. Although this approach elevated the Soviet Union to the position of one of the world's two superpowers, this was not sustainable and ultimately contributed to the country's demise (Snyder 1987/8: 107).

The importance of a strong military in Russia's self-perception as a great power provides important context for a nuanced understanding of recent developments. Given that a powerful military has always been seen as a necessity in this respect, the recent military revival is not a surprise. For Russia, military power has always been the means of choice for bridging the gap between its self-perception as a great power and what it viewed as the reluctance of other global powers to grant it this status. In part, this means that the recent military revival has already achieved a central foreign policy objective, as Russia is yet again seen as a force to be reckoned with. This achievement is a double-edged sword, however. As was the case with the Kremlin's use of energy as a tool of foreign policy, the display of military power in Crimea and Syria has not resulted in a positive image. Moreover, evidence from history suggests that the pursuit of great power status based on military might alone has never been sustainable. Whether the recent military revival will lead to a different result is far from certain. As Fyodor Lukyanov acknowledged in 2016:

> There is no doubt that during the past few years, Moscow has achieved some successes in its quest to regain international stature. But it's difficult to say whether these gains will prove lasting. The Kremlin may have outmanoeuvred its Western rivals in some ways during the crises in Ukraine and Syria, [. . .] but Moscow's failure to develop a coherent economic strategy threatens the long-term sustainability of its newly restored status.

Sovereignty

Sovereignty has been the central principle of international society since the late seventeenth century. As Charles Ziegler has written: 'key to the definition of sovereignty is the exercise of authority over a geographically defined territory' (2012: 401). Sovereignty has an internal and external dimension. As Robert Jackson put it, 'Sovereignty, strictly speaking, is a legal institution that authenticates a political order based on independent states whose governments are the principal authorities both domestically and internationally' (1999: 432). The defence of territorial jurisdiction and sovereignty to conduct its internal and external affairs without outside interference has been central to Russian foreign policy, as is the case for most states, throughout its history. However, since the collapse of the Soviet Union, which presented Russia with a serious crisis of statehood and identity, the importance of maintaining the country's sovereignty has emerged as a key principle and become, as Viatcheslav Morozov has written, 'the top priority in the Russian foreign policy agenda' (2010: 2).

Throughout his presidency, Putin has stressed Russia's need to protect its sovereignty on many occasions, for example, in his annual State of the Nation speech in 2014. In this speech, he asserted that 'true sovereignty for Russia is absolutely necessary for survival' (Putin 2014b). To outside observers, Russia's ongoing concern about sovereignty might appear surprising, because international attention has tended to focus on problems pertaining to the sovereignty of the other former Soviet states in view of Russian efforts to control or dominate them. Moreover, the international community has never seriously doubted Russia's status as a sovereign state with exclusive jurisdiction over its territory or political process. In order to understand the centrality of sovereignty in Russian foreign policy discourse, it is important to understand the complexity of the Kremlin's understanding of what 'true' sovereignty denotes, why it thinks it is threatened, and what policies it has adopted to counter these perceived threats.

Before discussing the specificities of Russian views on sovereignty, especially in the post-Soviet years, it is important to note that, like most states, Russia perceives a strong military as an absolute requirement for the preservation of sovereignty on the most fundamental level. Having the capacity to defend a state's territorial integrity has, historically, been viewed as an essential requirement for ensuring sovereignty (Rudolph 2005: 7–8). The feeling of vulnerability, in Russia's case, has been aggravated throughout the country's history by its size, long borders and geopolitical position. In Suny's view, geography constitutes a major factor that can explain certain trends in Russian foreign policy: 'Russia today, as in much of the past, lives in a dangerous neighbourhood. Both real and perceived dangers have historically contributed to its sense of weakness and vulnerability' (2007: 35–6). Not surprisingly, all Russian military doctrines issued since the end of the Cold War, which are discussed in more detail in chapter 5, list the defence of Russia's sovereignty and territorial integrity as the first task of the armed forces. In Putin's view, military power is fundamental to the country's sovereignty. Military modernization was therefore inevitable, as he noted in 2006: 'we need armed forces able to simultaneously fight in global, regional and – if necessary – also in several local conflicts. We need armed forces that guarantee Russia's security and territorial integrity no matter what the scenario' (Putin 2006).

That said, sovereignty is a contested concept. Its meaning has varied throughout history and has been interpreted differently by individual states (Barkin and Cronin 1994: 108–9). As such, in order to understand why sovereignty is so central to Russian foreign policy, Moscow's specific understanding of the concept needs to be considered. Russia continues to adhere to a 'traditional' Westphalian reading of sovereignty. An aspect of Westphalian sovereignty particularly important to Russia is the principle of balance of power, which would prevent any one state from seeking hegemony in the international system (Jackson 1999: 441). Throughout history, Russia played an important role in maintaining a balance of power in Europe, not least because it had the military strength to do so. This included Russia's role in preventing Napoleonic France's bid for hegemony, Nazi Germany's expansion leading to the Second World War, and also the rivalry with the United States in a bipolar contest during the Cold War. In the post-Soviet years, Moscow's strong concerns about what it sees as a unipolar world order dominated by the United States became apparent already during Yeltsin's first term as president. Moscow came to understand that its loss of great power status meant that it no longer had the clout to shape developments of global importance. Russia's inability to prevent NATO's military operation against Slobodan Milosevic's regime in 1999, which it vehemently opposed, cemented this perception. As Charles Ziegler put it, 'the problem from Moscow's perspective is that Washington expects Russia to subordinate itself to the US-dominated international hierarchy that emerged after 1991. Russian leaders vehemently reject the implication that they should accept a subordinate international status within this new order' (2012: 412).

Russia's understanding that its sovereignty is threatened in this way shaped its military policies. For the Kremlin, it became increasingly clear that a weak military curtailed its freedom of international action as a side-effect of the loss of great power status. From 2000 onwards, strengthening Russia's military power became a priority as this seemed indispensable to protect the country's sovereignty to act as an independent pole in international politics, whose voice could not be ignored. As Putin summed it up in 2012:

the basic principles of international law are being degraded and eroded, especially in terms of international security. Under these circumstances, Russia cannot fall back on diplomatic and economic methods alone to settle contradiction and resolve conflict. Our country faces the task of developing its military potential as part of a deterrence strategy and at a sufficient level. Its armed forces, special services and other security-related agencies should be prepared for quick and effective responses to new challenges. This is an indispensable condition for Russia to feel secure and for our partners to heed our country's arguments in various international formats. (Putin 2012b)

Recent displays of the country's revived military power have had as one of their main objectives to force 'partners' to listen to Russia's arguments. As Fyodor Lukyanov put it, 'by taking action in Ukraine and Syria, Russia has made clear its intention to restore its status as a major international player' (2016).

It is no coincidence that Putin's statement quoted above notes the need to strengthen not only the regular armed forces, but also the special services and other security-related agencies, which are predominantly responsible for dealing with domestic order and stability. This is because Russia's view of sovereignty is characterized by a distinctly intertwined nature of its internal and external dimensions. The Kremlin believes that its sovereignty to conduct its internal affairs without outside interference can only be preserved if it can also pursue an independent foreign policy abroad. As Ziegler described this, Russian policy shows a 'close linkage between the recentralizing project domestically, and the reassertion of Russia's position as a great power on the international scene' (2012: 401). Russia has long regarded international adherence to the principle of non-interference in the domestic affairs of other states, which is central to the 'traditional' Westphalian view of sovereignty, as the key to protecting its freedom of action at home. Since the end of the Cold War, a general shift in views of security from a state-centred focus towards a more human-centred interpretation, meant that this principle as an absolute has come into question. In cases where states themselves present a threat to their citizens, state sovereignty is no longer always seen as a barrier to outside intervention to enforce compliance with humanitarian norms (Thomas and Tow 2002: 180). This view was enshrined as a new principle in international law, the Responsibility to Protect, by the UN in 2005 (Bellamy 2009).

Russia has supported interventions with a humanitarian remit in certain cases (Averre and Davies 2015: 823). However, the belief that the West is using such norms as a pretext to get rid of inconvenient regimes and to spread its own influence has become engrained and is seen as a serious threat to the country's sovereignty. This belief dates back to the Kosovo War in 1999, when Moscow interpreted NATO's Operation Allied Force as an act of unilateral aggression against one of its allies. As Yeltsin noted in an address to the OSCE in 1999, European security was endangered by 'calls for "humanitarian intervention" – a new idea – in the international affairs of another State, even when they are made under the pretext of defending human rights and freedoms' (Yeltsin 1999: 132). US-led interventions leading to regime change in Iraq in 2003 and in Libya in 2011 were seen as further evidence of this. As Putin put it in 2012, 'armed conflicts started under the pretext of humanitarian goals [are] undermining the time-honoured principle of state sovereignty, creating a void in the moral and legal implications of international relations' (Putin 2012a). The Kremlin's belief in the West's intent to expand its power by interfering in the internal affairs of other states has also informed its suspicions over Western support of the 'colour revolutions' in the CIS region and, ultimately, over civil society projects in Russia itself that are operating with outside funding (Averre and Davies 2015: 826). All of this has been interpreted as part of a wider plan of expanding the West's influence, if required by military force, and as such as a threat to Russian sovereignty.

Internationally, a revived military has enabled Russia to stand up to what it sees as Western efforts to undermine the principle of sovereignty as the basis of international order. Having been unable to prevent the forceful deposition of the Serbian government during the Kosovo War in 1999 as it saw it, a stronger military enabled the Kremlin to prevent a similar scenario in Syria in 2015. At home, concerns over sovereignty have led to growing centralization and state control over all aspects of society. This has included restrictions on media freedom and civil society through registration laws and other means (Bacon et al. 2006). It also has led to a process by which those force structures tasked predominantly with internal security and public order have been increasingly strengthened. This will be discussed at length in chapter 3.

Finally, it is important to note that 'Westphalian' sovereignty, at least as Russia subscribes to it, sees sovereignty as an absolute right of great powers that does not necessarily accord the same right to lesser powers within their sphere of influence (Deyermond 2016: 958). This explains Moscow's seemingly contradictory readiness to pursue interventionist foreign policies towards the 'near abroad', which has left 'Russia open to charges of hypocrisy and double standards', as Graeme Herd has remarked (2010: 26).

Imperial legacy

Imperialism, including imperialist expansion, have featured variously in Russian foreign policy throughout the country's eventful history. The annexation of Crimea raised suspicions about the (re)awakening of Russian imperialism and expansionism. Although Russia's imperial legacy is still a factor in its foreign policy today, the meaning of this factor in its contemporary form is far from straightforward.

The Russian Empire, from the seventeenth century until its demise as a result of the Bolshevik revolution, was an empire in the classical sense of the word. An empire, according to Ronald Suny, can be described as 'a polity based on conquest, difference between the ruling institution and its subjects, and the subordination of periphery to the imperial center' (2012: 21). As with all great empires of the past, the Russian Empire was created on the basis of expansion through territorial conquest. Expansion was pursued to avert potential threats by rival powers, but it also gave Russia access to valuable natural resources and granted it international prestige as a great power. The problematic belief that expansion was the only means to ensure the state's security shaped the foreign policies of many empires in the past (Snyder 1991: 1). The Russian Empire, at various points of its existence, was no exception. As the historian William Fuller wrote, 'throughout the seventeenth and eighteenth centuries, Russia entertained vast ambitions. In an age in which the only choice was to conquer or to be conquered, Russia wished to neutralize its three most dangerous enemies – Sweden, Poland and Turkey' (1992: 435). Although dissenting voices on the merits of expansion were not absent in the Russian Empire's history (Jones 1984), for most of the seventeenth, eighteenth and nineteenth centuries the glory of Russia's tsars, empresses, and generals was measured by the territory they conquered.

There was another side to the coin of territorial expansion as it also created new threats, challenges and possible costs. Expansion increased the number of potential foreign enemies and absorbed large swathes of various ethnic and religious minorities into the Empire. This meant that in the event of war, Russia ran the risk of fighting against a foreign enemy at the same time as combating an internal insurrection (Fuller 1992: 52).

The Russian Empire came to an end with the Russian revolution in 1917. The early Soviet state was intended to be temporary and transitory in the immediate post-imperial years and there were expectations that the Soviet Union would ultimately serve as 'an example of equitable, non-exploitative relations among nations' (Suny 2007: 48). Although anti-imperialist rhetoric regularly featured in official Soviet discourse until the country's collapse, ironically, as Suny noted, 'almost from its inception the Soviet Union replicated imperialist relations' (2007: 48). The foreign policy of the Soviet Union is often described as imperialistic. However, expansion through the conquest of territory did not figure in Soviet foreign policy making after the incorporation of the Baltic States into the Union during the Second World War and the addition of some further territories to existing republics immediately after the War. In this sense, imperialism here is used as a metaphor for external relations with neighbouring states, lack of legitimacy and instability (Suny 2012: 23). Imperialism is often associated with expansion, but the two do not necessarily always go hand-in-hand. In a similar vein to Suny's definition, Johan Galtung's influential account characterized imperialism as a 'special type of dominance of one collectivity, usually a nation, over another', where domination can be political, economic or cultural (1971: 116–17). It is in this sense that descriptions of Russia's conduct today towards the other former Soviet states as imperialistic, or post-imperialistic as Dmitri Trenin has called it, should also be understood (Trenin 2011; H. Smith 2016).

Post-Soviet Russian foreign and security policy has often been described as imperialistic since the early 1990s, especially regarding its relations with other states in the former Soviet region with the exception of Estonia, Latvia and Lithuania. The Baltic States evaded the potential for continuing Russian influence, first by opting out of the Commonwealth of Independent States in 1991, and later by becoming members of NATO and the European Union. For the rest, just as Russia had expected that it would automatically inherit the Soviet Union's great power status, it was also assumed that ongoing influence over the former Soviet region was a given. Already in 1992, the chair of the Russian Supreme Soviet's Committee for Foreign Affairs and Foreign Economic Relations, Evgenii Ambartsumov, wrote that:

as the internationally recognized legal successor to the USSR, the [Russian Federation] must proceed in its foreign policy from a doctrine declaring all the geopolitical space of the former Union as a sphere of its vital interests . . . and must seek the world community's understanding and recognition of its [special] interests in this space. Russia must also seek from the world community the role of political and military guarantor of stability on the whole former space of the USSR. (Page 1994: 794)

As it turned out, and unexpectedly for Russia, neither its neighbours nor the West shared this expectation. As the newly independent states developed their own foreign and security policies, they cooperated with Russia when it suited them, but also kept an open mind to other options. The West believed that, as sovereign states, Russia's neighbours were free to pursue their own interests (Renz and Smith 2016: 17–18).

Why is influence and a dominant position in the former Soviet region still so important to Russia? As subsequent chapters will show, insecurity and the perceived need for a buffer zone play a role. Given that Russia does not have a strong reserve of historically close allies, adding to its feelings of insecurity, having a 'sphere of influence' is particularly significant (Renz and Smith 2016: 16). Strategic and material interests, as well as status concerns, are also important factors. Military force has been a vital means for the preservation of Russia's privileged position in its perceived sphere of influence, starting with the various 'peacekeeping' operations in the early 1990s, as discussed at length in chapter 4. However, until 2014, Russian efforts to ensure its dominance and privileged position in the region never relied on territorial expansion or direct political subordination. Whether the annexation of Crimea was a reversal of this trend, or an exceptional case, is far from certain. As examples in Russia's own history show, the cost of expansion is significant if the suppression of resistant populations is involved, and prohibitive in situations that could lead to military confrontation with other great powers.

Multilateralism

Based on Russia's history and the importance its leaders have traditionally attached to a strong military, Western observers have often viewed it as an actor that prioritizes the maximization of its own power at the expense of existing institutions (Tsygankov 2009: 51). Recent developments in Russian foreign policy, and especially the operations in Ukraine and Syria, have led to renewed fears that the country is yet again preparing to go it alone in its quest for great power status. Such an interpretation of Russian foreign policy risks being one-dimensional, because it neglects the important role multilateralism historically has played in the country's international politics.

Russia's search for multilateralism dates back to the time of Alexander I. A new European security order – the Concert of Europe – was agreed in 1814–15 by four great powers: Russia, Prussia, Austria-Hungary and Great Britain. Tsar Alexander I and the British foreign minister, Viscount Castlereagh, both believed that regular meetings and discussions between rulers and foreign ministers under this framework would serve to maintain peace in Europe. Indeed, the Congress of Vienna treaty succeeded in preserving peace in Europe until the outbreak of the Crimean War (1853–5). For Alexander I, the Russian Empire's interests in Europe went beyond the expansion of territory and the maximization of power. He wanted Russia to play a leading role, together with other great powers, in the affairs of the continent and also beyond (Seton-Watson 1989: 174–5). This period of history is often referred to in Russia today as providing an example of the best way of ensuring international security and as a model for the country's desired global role (Baunov 2015). As Jeffrey Mankoff has put it 'The Russian worldview is analogous perhaps to the Concert of Europe' (2007: 129).

A similar system emerged after the end of the Second World War. The victorious powers agreed to respect each other's 'sphere of influence' and set up the United Nations. As permanent members of the UN Security Council, the victorious powers were granted a privileged place in this new institution. This new system of multilateralism operated at first on the basis of cooperation, and later on opposition between the great powers as the Cold War developed. While multilateral institutions remained, Soviet leader Nikita Khrushchev preferred bilateral relations between the world's two superpowers. This approach continued into the Brezhnev period, although multilateralism enjoyed a renaissance at the time of the 1975 Helsinki Final Act. Russia's search for multilateral engagement continued into the post-Soviet period. As Elana Wilson Rowe and Stina Torjesen wrote, 'both Boris Yeltsin and Vladimir Putin consistently professed a deep attachment to the principles of multilateralism' (2009: 1). In a long article discussing Russian foreign policy from a historical perspective in 2016, Foreign Minister Sergei Lavrov outlined his country's long history of cooperation through multilateral effort and stressed that this factor remained important. In his words, Russia had never been 'fighting against someone but for the resolution of all issues in an equal and mutually respectful manner as the only reliable basis for a long-term improvement in international relations' (2016).

Assessing the specific understanding of multilateralism in contemporary Russia is important for evaluating its role in the country's foreign policy, and statements like Lavrov's. This understanding, as noted above, remains heavily linked to nineteenth-century ideas about cooperation in Europe and, as such, to Russia's self-perception as a great power. Moscow's view on the concept has been described as 'great power multilateralism involving leading states that may or may not take into consideration the concerns and wishes of smaller states' (Wilson Rowe and Torjesen 2009: 2). In the sense that multilateralism is seen as an activity between great powers, it is closely related to the idea of multipolarity. In the Russian context, both concepts are often used interchangeably or in overlapping ways (Tsygankov 2009). Former foreign minister Igor Ivanov's book on Russian foreign policy reflected the same understanding of multilateralism. Citing examples of constructive interaction by permanent members of the UN Security Council, he argued that Russia did not pursue the vision of a multipolar world where it was in opposition to the West. Instead, he suggested that Moscow wanted to create a mechanism enabling collective responses to international challenges. In his vision, such a mechanism would enable the solution of international problems without compromising the national interests of any of the leading powers. It would also serve as a forum for the resolution of disputes between the great powers themselves (Ivanov 2002: 47). In effect, he proposed a club or 'concert', where great powers could cooperate, even if they continued to compete with each other in a multipolar setting.

This understanding of multilateralism has shaped Russia's involvement in multilateral security cooperation throughout the post-Soviet years. As multilateralism is seen exclusively in the context of multipolarity, cooperation has often been informal and with specific partners when joint security interests demanded it (Wilson Rowe and Torjesen 2009: 2). The Kremlin's support of the US-led global war on terrorism is an example of such informal cooperation, as discussed in more detail in chapter 4. However, Russia has also engaged in multilateral security cooperation within the framework of permanent structures. Russian contributions to UN efforts to deal with 'new' security challenges, such as natural or man-made disasters, have been numerous and successful. Russia's most sizeable contribution to UN peacekeeping was in the Balkans from the 1990s until 2003. This cooperation was in many ways successful, as discussed in chapter 4. However, as a result of tensions with NATO over Operation Allied Force, Russia's subsequent involvement in 'traditional' peacekeeping has been limited. There has been a particular reluctance to engage in operations led by Western institutions, and by NATO in particular, because the loss of independence of Russian troops within such a framework was not deemed acceptable (Adomeit 2009: 102).

Having said this, Russia has engaged in multilateral security cooperation with NATO under the auspices of the NATO–Russia Council. Although Moscow's view that it was not being treated by NATO as an equal partner meant that such cooperation was never free from friction, Russian military personnel have worked alongside the alliance in the areas of emergency response, disaster management and counterdrug operations. Following the annexation of Crimea, military-to-military cooperation involving Russian personnel was suspended. Russian cooperation both with the UN and with NATO to counter 'new security challenges', often involving the force structures, rather than the regular armed forces, is assessed in more detail in chapters 3 and 4.

Russia has also pursued multilateral security cooperation within formal settings under the aegis of the Commonwealth of Independent States (CIS), the Collective Security Treaty Organisation (CSTO) and the Shanghai Cooperation Organisation (SCO). The CIS was created in 1991 as a framework for dealing with the multitude of challenges arising from the division of the Soviet Union into 15 newly independent states. By 1993, all former Soviet republics, with the exception of the Baltic States, had joined as members. However, views on the purpose of the organization quickly started to diverge. Some of its members, including Russia, hoped that it would lead to integration and close cooperation, including in the security realm. Others saw it merely as a means for 'civilized divorce' and were suspicious that it could turn into an instrument of Russian domination. A number of Russian 'peace-keeping' operations in the CIS region during the 1990s were conducted under a CIS mandate. In 1993, Russia also tried to formalize peace-keeping and conflict resolution as functions of the CIS permanent structures. However, this initiative did not receive much support from other members. As the commonwealth countries' foreign policy priorities started to diverge, and many were wary of Russia's intentions, the organization's potential as a multilateral institution to provide security in the region was never realized (Kubicek 2009).

Membership in the CSTO, which was established in 2002 on the basis of the Collective Security Treaty signed in 1994, includes, in addition to Russia, Armenia, Belarus, Kazakhstan, Kyrgyzstan and Tajikistan.[1] As the latter states are all broadly allied to Russia, security cooperation within this framework has been more successful. The CSTO has been holding regular military exercises and maintains a Joint Rapid Reaction Force. An important area of security cooperation within the CSTO are 'new security challenges', such as drug trafficking, organized crime and terrorism. The CSTO officially proposed the establishment of inter-institutional cooperation on similar issues to NATO at various points since the mid-2000s, but was never successful with this request (Nikitina 2012: 46). Owing to the relative

---

1 Azerbaijan, Georgia and Uzbekistan were also signatories of the treaty, but later left the organization.

weakness of the other CSTO members, Russia's position within the organization resembles that of the Soviet Union in the Warsaw Pact. The CSTO secures the military allegiance of its members to Russia and contributes to the maintenance of its dominant position in the region (Torjesen 2009: 182). In this sense, it serves as an 'institutional framework for "balance of power multilateralism"', as Andrey Makarychev and Viatcheslav Morozov have argued (2011: 363).

The SCO was established in 2001 on the basis of the Shanghai Forum – regular meetings held between Russia, China, Kazakhstan, Kyrgyzstan and Tajikistan since 1996. Uzbekistan joined in 2001 and both Pakistan and India were admitted as members in 2017. Initially intended as a framework for confidence-building in the region, the SCO subsequently developed into the most important security organization in Central Asia (Aris 2009: 457). Although security cooperation within this framework has focused predominantly on issues like counter-terrorism and border security, various military exercises have also been held since 2002 (Torjesen 2009: 188). The SCO serves as an important forum for Russia and China – two major powers with a stake in Central Asia – to address common security issues and deconflict potentially competing interests in the region. It has been noted that Russian efforts to engage in multilateral security and military cooperation within the framework of the CSTO and SCO were stepped up as Russia's opposition to NATO eastwards enlargement grew, especially following the stationing of US troops in Central Asia in support of the war in Afghanistan (Legvold 2009: 35). This is consistent with Russia's close association of multilateralism with multipolarity. As Makarychev and Morozov wrote, Russian security and military multilateralism in the former Soviet region 'can be described as strategic use of Russia's influence as the former imperial centre with a view toward creating a counterbalance to the West' (2011: 362).

The case of the OSCE (Organisation for Security and Cooperation in Europe) differs from the clearly 'harder' security-focused frameworks discussed above. It has also been a framework where Russia and the West have generally failed to find common ground, even if it is seen by Moscow as the best multilateral format for dealing with common European security concerns (Godzimirski 2009: 123). Russia sought to cooperate with the OSCE since the early 1990s and has made efforts to promote it as a multilateral security framework for the whole of Europe. Moscow was also keen to involve the OSCE in finding a solution to the conflicts in Transnistria, South Ossetia, Abkhazia and Chechnya (Gorzimirski 2009: 129–30). However, dissonant views on multilateralism, where the OSCE emphasizes equality and Russia insists on the primacy of great powers, has created tensions and stalemate in many cases in the past. That said, in spite of Russia's much more critical view of the OSCE today, it was still willing to allow a place for the OSCE in the war in Donbas and has argued strongly for multilateral conflict resolution in the 'Normandy format' – negotiations involving Germany, France, Ukraine and Russia.

In sum, the role of multilateralism in Russian foreign policy is complex, but it has been important throughout the country's history. Moscow's views on multilateralism differ significantly from those in the West. Instead of seeing it as a 'horizontal tool' that gives an equal say to big and small powers alike, the Kremlin advocates 'great-power multilateralism', where decisions are the prerogative of leading states. Multilateralism and cooperation, including in the security realm, is important to Russia, because true great power recognition can only be achieved within the framework of interaction with other leading powers. However, in multilateral settings where Russia does not feel like it is on an equal footing with other great powers, this can lead to conflict and validates the belief held by some in the West that Moscow is seeking to maximize its power at the expense of international institutions.

Conclusion

This chapter assessed the role of military power in Russian foreign policy throughout history. Focusing on the factors of great power status, sovereignty, imperial legacy and multilateralism, it provided essential context and background for the arguments presented in the following chapters. The chapter showed that having a strong military has always been important to Russia. The reasons for this go far beyond the desire to fight offensive wars, to expand territory, or to push for global domination. This is because Moscow has always seen military power as a flexible tool of foreign policy. An appreciation of the various roles the military has played in the country's past is important for an informed understanding of the reasons for and implications of the recent military revival.

A powerful military has always been of symbolic importance for Russia and an essential element in its self-perception as a great power. Past lessons have taught the Kremlin that the country will only be accepted as an equal by other leading powers, if it is seen as a force to be reckoned with. As the next chapter shows, status concerns have informed reforms of the Russian armed forces since the early 1990s. However, lessons from the past, which show that strengthening the military without being able to compete in other ways often leads to failure, are still relevant today. The prospects of catching up fully with the West's military capabilities remain in doubt.

Sizeable and capable armed forces have also always been essential for Russia's territorial defence. The country's vastness, long borders and historical exposure to foreign attack have made this factor particularly important. This is still the case today and has been a central reason for the military revival, as all of the remaining chapters will variously demonstrate. Russia's concerns over sovereignty are not limited to territorial integrity, as chapters 3, 4 and 5 explain further. Military power is also required to ensure the country's freedom of action internationally, as well as to protect order and stability at home.

Military force was essential for the creation of the Russian Empire and used for territorial expansion during the seventeenth, eighteenth and nineteenth centuries. Although the country's imperial legacy continues to be a factor in foreign policy, this has predominantly expressed itself in more indirect forms of control over neighbouring lands over the past century. As chapter 4 will detail, imperialist sentiments have informed Russian decisions to use military force in the CIS region on numerous occasions since the early 1990s. The chapter will also explain why the re-emergence of an expansionist vision is at least highly unlikely.

Finally, rather than seeing the military, above all, as an instrument for confrontation, the Kremlin has used it also as a tool for multilateral engagement at various points of its history. As chapters 3 and 4 will show, Russia has cooperated with international institutions and individual states towards the solution of various security challenges throughout the post-Soviet years. The desire for inclusion remains a vital factor in Russian foreign policy. An understanding of this is important, especially in times of heightened tensions, like in the aftermath of the annexation of Crimea in 2014. However, the Kremlin's specific view on multilateralism, which sees this largely within the context of multipolarity, can lead to conflict in situations where it feels that it is not being treated as an equal partner.

With the context of the role of the Russian military in foreign policy throughout history in mind, the following chapters assess reforms of the armed forces since the early 1990s, the significance of the force structures, the Kremlin's use of military force since the collapse of the Soviet Union, and changes in strategic thought. A detailed understanding of these issues allows for an informed assessment of the implications of the military revival for Russia's neighbours and for the West.

Chapter 2

Reforming the military

The break-up of the Soviet Union in 1991 went hand in hand with the demise of the once-powerful Soviet military. The newly independent Russian Federation kept the bulk of the Soviet armed forces' manpower and assets. However, as discussed in the previous chapter, the expectation that, as the heir of the Soviet Union's nuclear deterrent and main beneficiary of the former superpower's material military might, Russia would automatically keep its international status as a global military power did not come to pass. In fact, when the Russian armed forces were created in 1992, it quickly emerged that the Soviet legacy was more of a curse than a blessing. Reform attempts throughout the 1990s summarily failed to turn the Russian military into a force fit for the twenty-first century. Although the country's nuclear deterrent was always maintained, it soon became clear that strong nuclear capabilities were insufficient for coping with the military challenges of the post-Cold War security environment as well as for upholding Russia's status as a great power. Russia's unreformed armed forces performed woefully when deployed to deal with the ethnic conflicts that had erupted across the former Soviet region in the early 1990s and suffered humiliating failures in Chechnya. By the end of the 1990s, Russia had largely been written off as a global military force as it was generally assumed that its armed forces stood 'perilously close to ruin' (Arbatov 1998: 83).

The Russian armed forces started to recover when a systematic programme of military modernization was announced in 2008. Underpinned by significant financial resources and political will to enforce its implementation, the programme resulted in substantial improvements in conventional military capabilities. These were demonstrated to the world during the military operation in Crimea in 2014 and the subsequent intervention in Syria, changing the international image of the Russian armed forces almost overnight and leading to a debate about Russia's military revival (Trenin 2016c).

This chapter shows that the perception of a sudden resurgence of the Russian armed forces and the assumed implications of this for international security require both historical and comparative contextualization. The restoration of Russian military power under Putin is best understood within the context of the various factors that led to the neglect of the armed forces and lack of fundamental reforms during the 1990s. These factors can by no means be reduced to fundamentally different views held by the Yeltsin and Putin leaderships on the importance of maintaining powerful armed forces and the status of a global military power. As the previous chapter showed, having a strong military has always been central in Russia's self-perception as a great power and the military has played various roles and functions in the country's foreign and defence policy throughout history. This did not fundamentally change after the end of the Cold War. Moreover, the extent of Russia's military revival should be measured not only against its capabilities during the 1990s. A state's military power, like its power as an actor in the international system as a whole, is always relative to the power of other states (Mearsheimer 1994/5: 10–11). Although improvements in absolute terms are certainly impressive, Russia's relative military power compared to other global military actors, especially with regard to its conventional capabilities, continues to be limited.

A time of troubles: post-Soviet reforms of the Russian military

The Russian armed forces underwent a drawn-out period of neglect and were left to fall into a state of serious disrepair during the Yeltsin years. In part, the lack of substantial reform in the early post-Soviet years occurred because transforming the armed forces simply was not highest on the list of priorities at the time. Russia not only had to build a national military from the remnants of the formerly powerful Soviet armed forces once it had become obvious in 1992 that the preservation of joint military forces covering the entire territory of the Commonwealth of Independent States was not a realistic option. The sudden dissolution of the Soviet Union meant that Russia had to undergo an all-encompassing transformation of the state. Preoccupied with the demands of urgent political, societal and economic problems, the country's leadership spent little time on rehabilitating the military in a systematic manner. It is also clear that in the tumultuous political times of the early post-Cold War years, Yeltsin prioritized the security of his own political power over the long-term security interests of the country as a whole (Vendil Pallin 2009: 68).

As discussed in the next chapter, rather than strengthening the armed forces as an institution, Yeltsin set out to fragment the Soviet security apparatus, including the Ministry of Defence, in order to avoid any one institution from becoming too powerful and potentially posing a threat to his nascent regime. At the same time, he strengthened individual force structures and security services other than the regular military that were seen as particularly loyal and whose tasks and functions appeared to be more relevant at a time when a clear and present danger from an external enemy was absent. A process of structured military reform was also hindered by the fact that the newly created armed forces of the Russian Federation were literally thrown in at the deep end. As detailed further in chapter 4, before any serious discussions about their future structure and outlook could take place, Russian units were deployed to deal simultaneously with several low-intensity conflicts across a range of former Soviet states and within the country's own territory.

It has to be borne in mind that creating modern and efficient armed forces from the remnants of the Soviet military was a monumental task that would have been difficult to achieve even under the most conducive circumstances. Russia inherited around 2.8 million servicemen from the Soviet armed forces, which had been maintained at a personnel level of over four million during the Cold War. Russia also took possession of large quantities of tanks, aircraft and other military equipment. Although the quantity of material and personnel assets Russia had at its disposal for the basis of a new national military force was impressive, much of this legacy was not suitable for the early post-Cold War conflicts, such as ethnic conflicts, peace operations, separatism and insurgencies, that Russia was engaged in. The Soviet army had been a mass mobilization military based on conscription that was configured and trained predominantly for high-intensity warfare in the European theatre (Bluth 1998: 75–6). Much of the equipment Russia inherited was obsolete. For geopolitical reasons, most of the best-equipped Soviet units and facilities, such as anti-aircraft units and airfields, had been stationed on the western and southern peripheries of the Soviet Union and were transferred to the national militaries of other former Soviet republics. Having lost many vital assets of the formerly integrated Soviet military structure – including command, control and communications systems, missile early-warning facilities and integrated logistical support – then-Minister of Defence Pavel

Grachev noted that Russia had inherited nothing more than 'ruins and debris' (Allison 1993: 28).

Military reform was not particularly high on the Russian political agenda, at least for the first decade of the post-Soviet era. It is also likely that the Russian leadership assumed that, with the threat of a large-scale conflict with the West greatly diminished, the sheer size of the conventional capabilities the country had inherited from the Soviet armed forces would be sufficient for dealing with small wars and insurgencies in its immediate neighbourhood. But the fact that military reform was not a top priority does not mean that there was no awareness that reforms were a necessity, at least in principle. Central components of the 2008 modernization programme, such as the need to professionalize and create rapid reaction forces, were discussed as early as 1992. However, although several substantial programmes for reorganization were announced throughout the 1990s, they failed to result in fundamental transformation (Renz 2010: 58). Political reasons for this failure, including the lack of willingness to see through changes unpopular with the military leadership as well as institutional infighting and lack of agreement on the armed forces' roles and missions, were significant and have been well documented by analysts throughout the post-Soviet years (Vendil Pallin 2009: 9–14).

In addition, the country's dire economic situation meant that ambitious plans for the Russian military were simply not realistic at the time. As is well known, the Russian economy was in serious trouble throughout the 1990s. Economic decline culminated in the devaluation of the rouble in 1998 and economic recovery occurred only from 2000 onwards, not least owing to the worldwide rise in gas and oil prices. Throughout the 1990s, public spending stalled, leading to a crisis in many state-funded institutions, including education and the healthcare system. The armed forces were no exception. In some ways, the military, which had enjoyed a priority position in the Soviet Union, where the pursuit of military might was a central driver of the economy, was particularly hard hit when it lost this privileged position after the end of the Cold War (Zatsepin 2012: 116).

Although the true extent of Soviet military expenditure is not known due to the secrecy surrounding these issues, the defence budget was vast and comprised around 10 per cent of GDP, with some estimates putting this figure at over 15 per cent towards the end of the Cold War (Cooper 1998; Harrison 2008). In 1992 this figure plummeted to 4.6 per cent and decreased further to its lowest point throughout the post-Soviet period at around 3 per cent in 1998. In real terms, this meant that the defence budget decreased from an estimated US$344 billion in 1988 to US$58 billion in 1992, reaching an all-time low of around US$19 billion per year in 1998 (SIPRI Military Expenditure Database).[2] Such drastic cuts in the funding of the armed forces throughout the 1990s meant that their capabilities and assets degraded due to insufficient money being spent on procurement and upkeep of equipment and training. The necessary withdrawal of Russian military personnel from Eastern Europe and other former Soviet republics and the decommissioning of both excess personnel and assets to reduce force levels were in themselves extremely costly processes. This further diverted scarce funds from efforts to transform the remnants of the former Soviet military into a more modern armed force (Allison 1993: 34).

From an equipment and capabilities point of view, the Russian air force and navy suffered most, because both are reliant on technology easily subject to degradation and their personnel require intensive specialist training. From 1993 until 2009, the air force received only

---

2 All figures on defence spending in this chapter are taken from the SIPRI Military Expenditure Database 1988–2015, accessed on 11 October 2016. Real-term expenditure is shown in constant US dollars (2014).

a handful of new SU-34 combat aircraft. Pilot skills suffered as short-ages of fuel and insufficient numbers of functioning aircraft meant that aircrew were able to fly only a fraction of the regulation training hours. The navy's share of the defence budget decreased dramatically throughout the 1990s and it only procured a single new large vessel during this decade, the missile cruiser *Petr Velikii*, in 1996. Apart from a number of ballistic-missile submarines, which were consistently maintained as part of the nuclear triad, Russia had no real operational navy until the 2008 modernization plans addressed this situation. The ability to project naval power beyond one's shores arguably is an essential characteristic of a global military actor. The degradation of this ability contributed significantly to Russia's loss of this status in the eyes of the world. Moreover, until around 2000, the Russian armed forces received no combat training to speak of because of the dearth of funding available. Joint large-scale exercises, enabling the training of combined arms operations, were introduced only in the first half of the 2000s, as were long-range patrols of TU-95 bombers, which had last been seen in 1992 (Herspring 2005a: 139; Barany 2007: 57–8; Vendil Pallin 2009: 99; Renz and Thornton 2012: 48–9).

A significant consequence of the lack of funding during the 1990s was the degradation of the image of military service as a profes-sion owing to steadily worsening service conditions. Salaries paid to Russian officers until the mid-2000s were far from competitive, even compared to those received by civilians working in the public sector. Poor living standards and the lack of adequate housing for military personnel were major concerns. Soldiers were routinely unable to access benefits they had been promised contractually and, in spite of the scores of political initiatives to deal with this situation throughout the 1990s, the problem was never resolved. This devalued the prestige and desirability that was afforded military careers during Soviet times and led to serious difficulties regarding the retention and recruitment of professional military personnel. It also aggravated the problem of corruption and crime within the military. The poor image of military service affected the willingness of young Russian men to serve as conscripts. Reasons for the low esteem of conscript service in Russia are well known. In addition to the fate suffered by many young draftees during the first Chechen War, the image problem was the result of poor conditions of service and in particular the notorious *dedovshchina*, a brutal practice of hazing and violence against soldiers with sometimes fatal results (Herspring 2005b). This further exacerbated the country's already serious difficulties with calling up enough conscripts as a result of the demographic crisis and declining birth rates. During the 1990s, the majority of men eligible for conscript service evaded the draft by way of exemptions or by paying a bribe. Inevitably, a lack of pride in their profession and feeling of humiliation amongst military personnel further degraded the armed forces' effectiveness and capabilities (Golts 2004; Renz 2012b: 200–2).

It is beyond doubt that the Russian military was subjected to serious neglect throughout the 1990s, especially compared to the status it had enjoyed throughout the Soviet era. However, the often hyperbolic portrayals of the Russian armed forces up until the annexation of Crimea as a military perilously close to collapse went too far. In spite of all the problems it experienced, the Russian military was able to address the country's perceived military security threats when it was tasked to do so by the political leadership. As discussed in chapter 4, Russian soldiers were deployed in large numbers to conflict situations both within and outside of the country's borders throughout the 1990s. In relative terms, the Russian military outrivalled that of any of the other former Soviet state at any point of the post-Cold War period, due to the sheer disparity in size and the fact that the militaries of those countries were

affected by similar levels of neglect. Although the operational performance of Russian forces in conflicts fought up until the war in Georgia
in 2008 was far from stellar, and especially the Chechen wars stretched
their capabilities in every possible way, the country never risked a situation that could lead to its forces' comprehensive defeat. It is also fair
to note that the failure to cope with ethnic conflicts and insurgencies is
not unique to the Russian military.

It is important to bear in mind that throughout the post-Soviet
period, a number of Russian quasi-military organizations made up for
some shortcomings in the regular armed forces' capabilities for dealing with military operations other than war. As discussed in detail in
the next chapter, Russia maintains a number of institutionally distinct
militarized force structures other than the regular army specializing
in specific small-scale contingencies and soft security threats. These
are rarely factored into assessments of Russian military capabilities,
because they do not fit neatly into a Western framework. However, they
have been making a significant, and often overlooked, contribution to
Russian crisis response, humanitarian missions, counter-terrorism
campaigns and anti-drug operations, both within the country and on
an international level.

Throughout the 1990s, the Russian military retained its capability
to deter potential global threats from further afield with one of the
world's strongest nuclear arsenals. Unlike the other branches of the
armed forces, the strategic rocket forces were maintained throughout
the post-Soviet period and, as a result, so too was parity in nuclear
military power with the United States. The Russian defence industry
held its position as one of the world's largest arms exporters throughout the post-Soviet era, even if this did not benefit its own armed forces
at the time. Although Russian defence producers have not been able
to compete with Western high-tech weaponry and electronics, they

have always been competitive in many other areas, including the production of fighter jets, tanks, helicopters and submarines, and have
exported such legacy systems in large numbers (Renz 2014: 65). There
is probably much truth in what Dmitri Trenin and Aleksei Malashenko
have called an old adage: 'the Russian army is never as strong as it
describes itself, but it is never as weak as it seems from the outside'
(2004: 112).

The period of neglect of the armed forces during the 1990s seriously
degraded Russian military capabilities and power. However, as discussed in the previous chapter, lack of attention paid by the country's
leadership to the maintenance of a strong military is uncharacteristic
if seen within the context of modern Russian history. In many ways
this period of neglect should have come as more of a surprise than
the more recent efforts to revive Russian military power. The Russian
armed forces experienced their 'time of trouble' throughout the Yeltsin
years for the combination of reasons outlined above. These reasons did
not include the conscious decision on the part of the political leadership to give up on the aspirations of being a global military actor, or
the belief that a strong military was no longer necessary. It is true that
when the Cold War ended, many believed, both in Russia and in the
West, that the centrality of military power would diminish, not least
because with the end of bipolarity the threat of a global conflict had
waned. However, such beliefs were short-lived when it emerged that
military power continued to be an essential instrument of statecraft
especially for great powers, such as the United States and also increasingly China (Renz 2016a: 24–5).

The understanding that a strong nuclear deterrent alone is insufficient to uphold the country's great power status when other countries'
conventional armed forces continued to modernize at a rapid pace,
long predated the Russian military's revival that began in 2008. The

first Russian military doctrine published in 1993 envisaged significant cuts to Soviet legacy force levels and prioritized the development of forces able to deal with local conflicts, which were the most immediate concern at the time. However, the idea that strong conventional military power was no longer desirable or required was never a consensus view. Many Russian military thinkers continued arguing in favour of more open-ended defence requirements that would keep the country prepared for a wide variety of eventualities (Arbatov 2000: 7). In fact, the 1993 doctrine already reflected serious ambitions to maintain strong conventional military power in addition to strengthening capabilities for dealing with small wars and low-intensity missions. It imagined investments in research and development (R&D) for the production of advanced weaponry and equipment, including electronic warfare capabilities and stealth technology. This was a direct response to the lessons Russian strategists had identified from the successes of the so-called Revolution in Military Affairs showcased by US conventional forces for the first time in the 1991 Gulf War (Pipes 1997: 75–6). At the time these plans were highly unrealistic and remained nothing but a pipe dream.

The ambition for parity in conventional military power was reiterated in the 2000 military doctrine, which explicitly reoriented priorities away from the focus on small-wars scenarios and towards the need for the creation of Russian conventional forces with global reach. This doctrine was issued in the wake of NATO's high-tech intervention, Operation Allied Force, over Serbia in 1999 which, as discussed in more detail in chapter 4 and in the words of Aleksei Arbatov, 'marked a watershed in Russia's assessment of its own military requirements and defense priorities' (2000: 8–9).

Russia's quest for great power status dates back centuries and its self-perception as such certainly did not cease with the end of the Cold

War. Military power was central to the making of the tsarist empire and it was also a strong military, above all else, which elevated the Soviet Union to the status of a superpower during the Cold War years. Relinquishing armed strength and accepting the resulting loss of great power status was never an option that was seriously entertained in Russia. From this point of view, the revival of the Russian military was only a matter of time.

The 2008 modernization programme and Russia's military revival

The Russian armed forces' fortunes started to change with Putin's appointment as Prime Minister in 1999 and election to the presidency the following spring, at a time when the second Chechen campaign was in full swing. As discussed in chapter 1, Putin afforded military-related matters more political importance from the outset, as he saw a strong and proud military as a prerequisite for Russia if the country was to regain international recognition as a great power and a strong, sovereign state. The need for military reforms was the subject of several discussions in the Security Council throughout 2000. In a speech to the country's top military commanders in late November 2000, Putin summarized these meetings' conclusions, emphasizing the urgent need for modernization in the areas of financial efficiency, discipline, combat readiness and available technology. Although he confirmed that the task of strategic deterrence had been 'successfully fulfilled', he pointed out that the Russian armed forces were not sufficiently prepared 'to neutralize and rebuff any armed conflict and aggression' that could come from 'all strategic directions', as demonstrated by the operations in Chechnya. Recognizing the 'hard work' of Russian soldiers in this conflict, he also criticized that the operations there had come at too high a cost and that the losses occurred were 'unpardonable'. In the same speech, Putin spoke about the need to improve the image of the military profession and to 'put an end to the humiliating situation of servicemen'. He asserted that ensuring soldiers' well-being and pride in their profession was as much a major reason for reforms as a requirement for their success. In his words, 'the problem is directly linked with national security interests. The trust of the army in the state, and having the army "feel good" about itself is the bedrock foundation of the state of the Armed Forces' (Putin 2000).

Finally, Putin noted that, although the need for military reform was urgent, it could not come at any cost and that the country 'should not just plan what we need, but plan proceeding from what we can afford' (Putin 2000). Assisted by a recovering economy and a GDP that experienced consistently positive growth rates from 1999 to 2008, not least due to rising oil and gas prices, the Russian defence budget increased from its low point of US$19 billion in 1998 to around US$58 billion by 2008, growing to more than US$90 billion by 2015. As discussed below, this growth was achieved without significantly raising the percentage of GDP compared to the Yeltsin years, at least initially.

Although plans for military transformation had been long in the making, the short war with Georgia in August 2008 served as a catalyst for the announcement of extensive military modernization in the autumn of the same year. In this war the Russian military had achieved strategic victory in merely five days, but its operational performance was again severely criticized both in Russia and abroad. In particular, there was widespread agreement that it still showed major shortcomings in coordination, command and control, as well as a lack of technology and weaponry fit for the twenty-first century (Bukkvoll 2009; Vendil Pallin and Westerlund 2009).

Continuing the direction envisaged by previous rounds of reforms, the 2008 modernization programme emphasized the efficiency of command structures, the need for more rapid reaction and the modernization of technology. As a whole, it was presented as a package that would allow the Russian armed forces to overcome the shortcomings they had experienced in Georgia and other previous military interventions and to finally do away with the Soviet legacy force (Klein 2012: 30). The programme sought to make the Russian military more useable by increasing its overall efficiency and cost-effectiveness: streamlining central command bodies; decreasing the size of the officer corps, which had made the Russian military particularly top-heavy; cutting the number of military units in favour of a smaller number with permanent readiness status; and driving up the recruitment of professional soldiers in order to lessen reliance on conscription (Sinovets and Renz 2015: 5). The image problem of the military profession was also tackled with improvements to the financial rewards and welfare of soldiers. As discussed in more detail below, a centrally important element of the modernization programme was the updating of weapons and equipment with a view to moving from a figure of 10 per cent of hardware classed as 'modern' in 2008 to 30 per cent by the end of 2015 and to 70 per cent by 2020.

Backed up by solid funding and unprecedented political will at the highest level, the complex reform programme, unlike previous attempts, was implemented with determination. The achievements of these 2008 reforms are well documented and there is widespread agreement that they have turned the Russian military into a force that is unrecognizable compared to the demoralized and underfunded organization it had developed into during the 1990s. Structural changes in particular were pushed through with impressive speed. The Soviet-era principle of the mass mobilization army meant that until

2008 the majority of Russian territorial units had been manned by only a skeleton staff of officers during peacetime, waiting to be filled out with mobilized reservists in the event of a major war. As a result, only 17 per cent of Russian military units were permanently staffed and ready to be deployed at short notice. This was not only inefficient, but also unsuitable for a security environment where military units have to be deployed to deal with various contingencies quickly and at short notice. In order to enhance the mobility and combat readiness of Russia's ground forces, they were reorganized from a four-tier command structure based on divisions to a three-tier command structure based on smaller brigades.

The rationale for this change was to increase the army's flexibility in creating more deployable units, simplifying the chain of command and enabling better coordination between the different arms of service during operations. In this sense, Russia followed a pattern of reform that had been pursued by many Western states, such as the United States and United Kingdom, which had adopted the brigade as the basic building block of its armed forces already during the 1990s (Renz and Thornton 2012: 47). As a part of this process, understaffed mobilization 'ghost' units were disbanded to leave room for a smaller number of units with permanent readiness. These structural changes were pushed through by December 2009. The process included significant cuts in the number of officers by about a third to around 220,000 by 2012. The rank of warrant officer was abolished completely. The reductions were achieved, in part, by slashing vacant positions, not replacing officers reaching retirement age and offering retraining opportunities to officers wishing to pursue a civilian career (Renz 2010: 58–9; Thornton 2013; Klein and Pester 2014).

The aim of increasing the number of professional servicemen was more difficult to achieve, because it required more than structural adjustments. However, advances in this area have also been made, because it was recognized that social aspects and improvements to the image of military service as a profession were essential if the process of modernization was to succeed. As Putin wrote in an article about the need for and achievements of military reforms in 2012, 'our aims in the sphere of defence and national security cannot be achieved unless . . . servicemen . . . are highly motivated – and unless, let me add, the Russian public shows respect for the Armed Forces and military service' (Putin 2012b). Competitive salaries, better service conditions and welfare provisions, including housing and pensions, improved the image of military service and left servicemen with a new sense of purpose and pride in their profession (Giles 2016: 16–17). Military service again became an attractive career option, particularly so in poorer Russian provinces and following the worsening economic situation from 2014 onwards. Measures were also taken to increase the appeal of conscription and to tackle the problem of *dedovshchina*. The term of conscript service was decreased from two years to twelve months. This was accompanied by a so-called programme of 'humanizing' service conditions for conscripts, including the introduction of regular periods of rest, improved nutrition, permission to leave the military unit over the weekend, and allowing personal calls from mobile phones when off-duty. As a result, draft dodging now is no longer a serious problem. Although Russia is still remote from the move to a fully professional military, the number of professional service personnel in the Russian armed forces increased from about 174,000 in 2011 to more than 300,000 in 2015 (Lavrov 2015).

Steps have been taken to make up for the dearth of new equipment procured by the armed forces throughout the Yeltsin era. The extremely ambitious plans for the updating of such equipment announced at the outset of the 2008 modernization programme were not achieved in their entirety for the reasons discussed further below. However, new equipment delivered as part of the state armaments programme to 2020 has undoubtedly made the Russian military more modern and more capable. Advances were made particularly in the realm of upgrading the strategic rocket forces, the country's air defence system and sizeable deliveries of new fixed-wing and rotary aircraft to the air force (Cooper 2016: 52). During the Crimea operation in March 2014, observers also noted the use by Russian troops of materiel that previously seemed unavailable, such as new 'webbing' and personal radios (Marcus 2014). In Syria, Russia demonstrated that its modernized military now had the sea and airlift capabilities required for limited out-of-area operations. This air campaign would have been beyond the realm of possibilities before the advances of the 2008 modernization programme took hold (Gorenburg 2016).

Structural changes and improvements in technology were accompanied by advances in education and training in order to enhance the mobility and combat readiness of the armed forces. Increased funding meant that large-scale exercises, which had not taken place for almost two decades after the end of the Cold War, were reintroduced in 2009 (Trenin 2016c: 24). Since 2011, inter-service exercises involving up to 150,000 men have been held on a regular basis, preparing all the armed services for joint and combined combat operations for the first time (Norberg 2015). Fostering jointness was a central goal of the 2008 modernization plans, because lack of coordination had been a serious problem in Russian military operations throughout the post-Soviet era. Although the Russian armed forces' capabilities in conducting sizeable inter-service operations have never been tested in an actual conflict situation, improvements in coordination, command and control were demonstrated in both Crimea and Syria on a limited scale.

The military modernization programme introduced in 2008 was the first plan for reform in the post-Soviet era that led to fundamental change. It is beyond doubt that it has resulted in considerable improvements in Russian military power. Having said this, there has been a tendency, especially in the West, to overstate the scale and implications of the reform programme both with regard to Russia's military posture in absolute terms and its relative standing as a global military power. Just as the Russian armed forces were never as close to collapse during the Yeltsin era as often asserted, the idea of a Russian military resurgence that has turned the country into a serious global threat within a few years is exaggerated. Although the ongoing modernization process will almost certainly lead to further improvements in capabilities, significant systemic obstacles continue to stand in the path of Russia's long-term military ambitions.

Manpower problems

Russia's struggle to maintain its desired level of military manpower is an ongoing problem with no simple solution. The Soviet Union's armed forces were comprised of over four million men. After the Cold War it was clear that near-Soviet force levels could and would not be maintained and significant cuts were required. The around 2.8 million personnel Russia inherited from the Soviet military were gradually reduced and a presidential decree issued in July 2016 fixed manpower levels in the armed forces at a maximum of one million (Presidential Decree 329, 2016). Although Russian officials have long referred to the one-million-man army, the country has struggled to achieve this level of manpower throughout the post-Soviet years. Russia's actual military strength today is usually estimated at around 800,000 men, with some analysts putting this figure as low as 625,000 (Carlsson et al. 2013: 38). Russia's desire to maintain a military of this size, in spite of not being able to do so, is often put down to conservative elements in the military leadership that have been unable to move on from Cold War thinking and who do 'not want to part with [their] massive army' (Barany 2005: 35). Throughout much of the 1990s and 2000s the consensus view was that the Russian military could never be fully modernized unless the large and mostly conscript-based army was abandoned in favour of smaller and more affordable professional units (Golts and Putnam 2004: 122; Klein 2009: 9). In the eyes of many observers, these would be fully sufficient for the 'small and "soft" security threats [Russia] should anticipate in the future' (Barany 2005: 49). The reasons for Russia's insistence on maintaining a military one million strong are more complex than the inability of conservative generals to move on.

It is clear that even a quarter of a century after the Cold War ended, quantity continues to matter as much as quality in the military ambitions of powerful states, including Russia. As John Mearsheimer put it, 'great powers require big armies' (2001: 6). Early post-Cold War expectations anticipating the saliency of military power to diminish and the idea that with the spread of democracy and economic interdependence, state competition between great powers in the future would revolve around economic matters, did not come to pass (Art 1996: 7–9). The maintenance of strong military forces for many states continued to be seen as an important asset. In the absence of a clear adversary against whom to measure required military capabilities, for example, the United States took the 'two-war' standard as the basis on which to decide the required size of its military in 1991. At this point in time there was no obvious existential threat emanating from a specific state adversary. However, it was decided that armed forces strong enough to cope with the potentiality of two simultaneous major regional wars were essential in order to ensure the country's ability to defend its territory and to deter potential foes, at the same time as maintaining capabilities sufficient for the engagement in crisis response, humanitarian operations and other contingencies (Goure 2013: 1).

Compared to the armed forces of other large powers, even a one-million-strong Russian military does not stand out. The US military has around 1.4 million professional personnel. China's People's Liberation Army is made up of a force of over 2.3 million professional servicemen and women (*The Military Balance* 2016: 484–6). Unlike the US, Russia does not have strong military allies and is unable to bolster its military strength with coalition warfare. NATO's total numerical strength amounts to over 3.3 million. A sizeable military is important for Russia's self-perception as a great power and for this reason it is unlikely that it will ever abandon quantity entirely in favour of smaller, more compact armed forces, even if this would make good financial sense. As discussed in chapter 1 and elaborated further in chapter 5, geostrategic concerns pertaining to Russia's territorial size and location are another factor why mass and at least a degree of mobilization have always been important. The perceived need to be prepared for a multitude of potential geostrategic threats from the South, East and West has been a constant in the history of the country's defence policy decision making dating back centuries (Kagan and Higham 2002: 2–3). In today's Russia, small, mobile forces would be fully sufficient for dealing with local conflicts, insurgencies and terrorism in the Caucasus, Central Asia and in other former Soviet states. Such forces would do nothing, however, to alleviate feelings of insecurity vis-à-vis potential future hostility from China and especially NATO, which returned to the centre of Russian threat perceptions during the 1990s.

Unlike the fully professional standing armed forces of the US and China, a large part of Russia's 800,000 service personnel continues to be made up of conscripts, further putting into perspective the limitations of its military strength. In 2008 the term of conscript service was reduced from two years to twelve months. Although this helped alleviate problems with draft evasion, the decision was also a double-edged sword. The shorter term in conscript service inevitably resulted in a drop in skill and experience levels, with many conscripts not even serving long enough to take part in a large-scale military exercise (Renz and Thornton 2012: 46). Owing to the high percentage of conscripts in the Russian armed forces, the implications of this for the quality of Russia's available military strength are considerable.

The reduction in the length of service also meant that now twice as many young men had to enter conscript service each year if manpower levels were to be maintained. By 2011, the annual intake of conscripts reportedly had stabilized at 300,000. Given that only about 700,000 Russian men reach conscription age each year, this number cannot be substantially increased in the foreseeable future. In other words, there are simply not enough young men to conscript in order to make up a military of one million (Lavrov 2015). The need to lessen the reliance on conscription and to professionalize the Russian armed forces has been a constant in all military reform initiatives from the early 1990s (Spivak and Pridemore 2004). A fully professional military continues to be on the wish-list of Russian military modernization, but statements by Defence Minister Sergei Shoigu and plans by his ministry indicated as late as 2013 that this would not happen at least until 2025 or even later (Lavrov 2015). Given the financial constraints of Russian military modernization (discussed further below), the move to a fully professional military is simply too costly.

Russia's inability to do away with conscription does not necessarily mean that its armed forces can never be fully modernized. The predominant tendency in the West has been to move towards fully professional forces over the past two decades, but some highly developed countries, including Israel, Norway and Finland, also continue to maintain a mixed recruitment system for their armed forces. In such systems, a core of highly trained professionals ensures the technological expertise and other skills required for the conduct of modern combined arms operations, whilst an element of conscription and mobilization is maintained for eventualities where greater numbers of infantry soldiers might be needed (Leander 2004: 572). In contemporary Russia, a system of combining conscripts with professionals in units is the only affordable option if a manpower level of even close to one million is to be maintained. This was recognized from the outset in the 2008 modernization programme. The former chief of the general staff, Nikolai Makarov, envisaged the creation of more permanent readiness units staffed entirely by professional soldiers. Such units would increase Russian rapid reaction capabilities within a mixed system of recruitment, which would also provide less-well-trained forces for use in 'calmer' areas (Klein 2009: 116).

The 2008 modernization programme has yielded results in this area. A growing number of professionals have been allocated to rapid reaction units in Russia's armed services. The airborne forces (VDV) benefited especially. Already containing fully professional elite regiments since 2002, the VDV came to have more than 50 per cent of its staff serving under contract by 2015. The percentage of professionals serving in specialist positions also grew, such as those involving the operation of advanced equipment and weaponry (Lavrov 2015). Advances in mobility were demonstrated in Crimea, where VDV units acted swiftly and in cooperation with other rapid reaction forces from the special forces reconnaissance brigades and naval infantry (Bartles and McDermott 2014: 57; Marcus 2014). These advances are certainly notable. It has to be kept in mind, however, that the elite troops who carried out the Crimea operation make up less than 1 per cent of Russia's overall military strength and are not representative of the Russian armed forces at large (House of Commons Defence Committee 2014a). A large proportion of this strength continues to be made up of poorly trained conscripts. Russia still has a long way to go to achieve the quantity and quality of fighting force it aspires to.

Economic stagnation

Financial limitations are a serious obstacle to the longer-term goals of Russian military modernization and for the ambitions to create a truly competitive conventional force. Since the early 2000s, defence spending has experienced consistent growth. Clearly, in absolute terms, the threefold increase in the Russian defence budget from below US$30 billion in 2000 to over US$90 billion by 2015 is dramatic, and similar developments during peacetime in a Western European setting are unthinkable. At the same time, the idea of a militarily resurgent Russia fuelled by a sudden increase in military spending requires contextualization, both internationally and in terms of the share of GDP Moscow has spent on defence throughout the post-Soviet period. In international comparative perspective, Russia's military revival is not as clear-cut as the rise in absolute spending on defence might suggest. It is obvious that defence spending alone is insufficient as a measure of a country's relative military power. Fluctuations in exchange rates and differences in purchasing power also complicate international comparisons in this area. Owing to the sheer discrepancy between Russian defence spending and that of other global military powers, this point is still significant. Russia's military expenditure of just over US$90 billion in 2015 was less than a sixth of US spending, which amounted to over

US$595 billion. Moscow spent less than half of the Chinese budget, which came to over US$214 billion in the same year.

To put the magnitude of Russian defence spending further into perspective, even when the country spent more than 5 per cent of its GDP on the military in 2015, its budget only slightly exceeded the combined military expenditure of Germany and Italy. Both countries do not have nuclear weapons and kept their defence spending comparatively low at 1.2 and 1.5 per cent of GDP respectively. Increases in spending elevated Russia to third place in the ranking of global military expenditure by 2015. The large gap between the two largest spenders and Russia, however, means that the latter continues to be in an entirely different league from the US and China and is much closer to the smaller countries in the mid-top ten of the list of global military expenditure. Russia's defence budget in 2015 was only marginally bigger than that of Saudi Arabia, the world's fourth-largest military spender. The UK and France, in fifth and sixth place, each had a budget of around US$60 billion, or approximately two-thirds of Russian spending. Russian military expenditure did not start exceeding that of France until 2011 and that of the UK until 2012.

Looking at Russian defence spending as a percentage of GDP, it is clear that the increase in military expenditure under Putin has not been all that dramatic. In fact, Russia's share of GDP expended on defence has been fairly consistent since the early 1990s. The average share of the GDP spent on defence under Yeltsin's leadership (1992–9) was 4.15 per cent, peaking at 4.9 per cent in 1994 and reaching a low point of 3 per cent following the economic crisis and devaluation of the rouble in 1998. The average share of GDP spent on defence under Putin and Medvedev (2000–15) was 4.16 per cent. This includes a sharp rise to 5.4 per cent in 2015 following the contraction of economic growth and the federal budget in previous years. The relative consistency in percentage of GDP spent on the military squares with the above argument that the slump in defence spending under Yeltsin resulted from economic weakness, rather than from a principled decision to afford less importance to a strong military. At the same time, it puts into perspective the idea of a sudden militarization of Russia since Putin's rise to political prominence. The revival of the Russian military is often associated with an about-turn in defence and security policies following the end of Yeltsin's years in office and the election of Putin as president in 2000. It is important to bear in mind that this revival not only coincided with Putin's appearance on the political scene, but also with a time of economic optimism and recovery.

A central factor in the recovery of Russia's economy was the sharp increase in the price of oil from an average of less than US$20 per barrel during the 1990s to almost US$150 by 2008. This heavily benefited Russia, because its economy continues to rely on energy exports, which account for about half of its federal budget income (Connolly 2015: 6). Between 2000 and 2008 the Russian economy grew by an average of 6.9 per cent per year, a rate of growth that was matched by similar annual increases in the defence budget (Oxenstierna 2016: 61). Following years of economic turmoil during the 1990s, when even a relatively high percentage of GDP spent on defence was insufficient to sustain the military on a reasonable level and made costly modernization unfeasible, Russia's long-standing ambition to rebuild a military fit for a great power had finally became affordable.

Affordability is an important factor in the future prospects of the modernization programme's longer-term goals that have not yet been achieved. The costly State Armament Programme 2011–25 unveiled in 2010 accounted for the major share of recent increases in defence expenditure. Its ambitious procurement plans, which are discussed in more detail below, had been based on the expectation of significant and consistent economic growth that would allow Russia to keep defence spending below 3 per cent of GDP for the lifetime of the programme (Cooper 2016: 51–42). These expectations did not come to pass, because Russian economic growth slowed as global oil prices steadily declined to below US$50 per barrel by 2015. The high levels of defence spending envisaged in 2010 became costly to the contracting economy and military expenditure ballooned to more than 5 per cent of GDP. The impact of the economic downturn has yet again presented the Russian leadership with the dilemma, as Michael Bradshaw and Richard Connolly have put it, of 'weighing the trade-off between spending on guns and butter' (2016: 156). This dilemma has been historically difficult for Russia to resolve, because its financial means rarely matched its military and great power ambitions.

As discussed in chapter 1, the Russian economy historically had to adjust to the needs of the military and not the other way around. Certainly, during Soviet times, a high level of military expenditure was affordable only at the expense of other areas of public spending or, as Vasily Zatsepin concluded, the 'guns-or-butter' dilemma was unambiguously resolved in favour of the guns (2012: 115). The pace and extent of further military modernization under the conditions of economic crisis will depend largely on the extent to which the Russian leadership is prepared to prioritize the defence sector above everything else yet again. By 2015 the evidence regarding its ultimate priorities and preparedness to pursue militarization at any cost was not yet conclusive.

The largely oil-induced economic downturn will inevitably affect the socio-economic situation of Russian citizens and put a strain on the social contract between the population and the regime in power, especially if populist promises made in the run-up to the 2012 presidential elections are not kept. As Bradshaw and Connolly have argued, by 2016 it was far from clear whether Russia could 'afford to continue to strengthen its military capability in the face of falling oil and gas exports revenues, economic recession, and growing social demands on the federal budget' (2016: 156). Russia continued to pay a comparatively high price for its military in terms of share of GDP, even though this percentage was lowered to a much less dramatic level after the collapse of the Soviet Union. When the share of GDP spent on defence started rising far above the previously intended level, this did not go unnoticed.

The process for the 2015–17 budget adopted in November 2014 was accompanied by 'fierce discussions' on the size of defence expenditure in the face of cuts to other areas of public spending. Finance Minister Anton Siluanov reportedly asserted that the country simply could not afford its current defence programme and had to adjust its ambitions to the changed economic situation. In the event, although some observers expected a cut in defence spending of around 10 per cent in the revised budget for 2015, this was reduced by only 5 per cent, as were allocations to budget items affecting the population most, such as health, education, culture and the environment (Oxenstierna 2016: 68–9). In the same year, approval of the next armament programme to 2025 was delayed for three years as decision makers appeared to be 'awaiting a more stable and predictable economic situation', as Julian Cooper put it (2016: 45). As mentioned above, Putin declared in 2000 that military reform could not come at any cost. In Susanne Oxenstierna's words, although the share of GDP allocated to the military remained high, in 2015 there was 'still a trade-off between defense and other spending in the budget' (2016: 68).

Defence industry and innovation

There are other factors limiting the Russian defence industry's ability to deliver 'modern' equipment to match the country's long-term military ambitions. After the Cold War, reforming the once-monolithic Soviet military-industrial complex was, in line with the reluctance to reform the military itself, not a priority of the Russian leadership. The Russian defence industry still, though, kept its position as one of the world's largest arms exporters throughout the post-Soviet era and remained competitive in the production of certain niche products, such as combat aircraft and submarines. However, little was done to bring the sector into the twenty-first century, because insufficient funds were reinvested in R&D (Renz and Thornton 2012: 50). As a result, the technology gap between Russian and Western producers continued to grow. Much has been written about the Russian defence industry's problems, which are said to include outdated management practices, a rapidly ageing workforce and an even older manufacturing base (Shlykov 2004; Cooper 2006; Blank 2012). These challenges did not escape the attention of the Russian leadership and modernization of the defence sector was made a priority when the military modernization programme was announced in 2008. It was recognized that the existing industry would not be able to fulfil the ambitious procurement targets of the State Armament Programme 2011–20, which was to modernize 30 per cent of the armed forces' equipment by 2015, rising to 70 per cent by 2020. A range of measures, such as increases in money allocated to R&D, more government defence orders and the allocation of additional resources to increase production capacity, were implemented and have yielded considerable results (Westerlund 2012: 90). In spite of the economic downturn from 2009 and delays in the delivery of some weapons systems, increases in procurement volumes since

2011 have been significant. By 2015 the State Armament Programme's accomplishments meant that the interim target of modernizing 30 per cent of military hardware was even exceeded in some areas (Cooper 2016: 51–2).

A closer look at the successes of Russian rearmament from 2011 until 2015 shows, however, that not everything was as rosy as it seemed. As Richard Connolly and Cecile Sendstad found in an in-depth study of the State Armament Programme, its performance by 2015 varied greatly across different sectors (2016). The industry performed best in areas where it had maintained production capabilities throughout the post-Soviet period. Targets were fulfilled mostly with the delivery of established equipment and systems based on upgraded legacy designs. At the same time, though, there was little progress made in the production of sophisticated high-tech weapons. As a result, the most successful implementation of the State Armament Programme has been in the strategic rocket forces (Cooper 2016: 47). This is not surprising, because in the area of strategic nuclear capabilities Russia had remained globally competitive even throughout the 1990s, when its other armed services were neglected. The modernization of the air force has also proceeded at a rapid pace, with more than 310 new aircraft delivered by 2015 out of 700 scheduled for 2020. However, as Connolly and Sendstad have found, this success was not unambiguous. The majority of new aircraft delivered to the air force by 2015 were front-line combat aircraft, such as the SU-34s, SU-35s and MiG-29K, and training aircraft, all upgraded legacy designs. The construction of entirely new models, such as the widely publicized PAK-FA fifth-generation fighter, has not proceeded to plan. Only five prototypes of the aircraft had been delivered by 2015 and some Russian analysts predicted that as few as twelve of the initially scheduled seventy could be supplied by 2020. There have also been difficulties with the production and delivery of strategic bombers and transport aircraft in sufficient numbers (Connolly and Sendstad 2016).

When it comes to the navy, targets were achieved in the delivery of upgraded versions of existing models of submarines and smaller surface vessels. The production of new models, such as the Yasen-class submarine and especially large surface vessels, experienced serious delays. The loss of input from Ukrainian defence producers and the US and EU economic sanctions imposed in the aftermath of the annexation of Crimea, including the cancelled delivery of French Mistral landing ships, seriously hindered the progress of naval modernization. The ground forces' inventory was modernized with large numbers of new armoured vehicles based on older models. The production of combat systems for infantry soldiers and artillery systems also proceeded to plan. Several hundred updated and modernized battle tanks were delivered, but only twenty prototypes of the new T-40 Armata main battle tank out of 2,300 scheduled for 2020 were produced (Connolly and Sendstad 2016).

In absolute terms, Russian military hardware today is incomparably more modern and technologically advanced than it was during the 1990s and much of the 2000s. In relative terms, however, the Russian armed forces are still a long way off achieving parity with the technologically advanced militaries in the West and the United States in particular. Since the start of the 2008 modernization programme, Russian defence analysts have cautioned that discussions of the desirable or existing proportion of 'modern' equipment in the armed forces were misleading, because the meaning of 'modern' equipment was too ambiguous. As Dmitry Gorenburg highlighted in 2012:

when Russian officials discuss their goals for procuring modernized weaponry over the next 10 years, they never define their terms. They do not have a list of what types of armaments are considered modern. In some cases, systems that are based on 20–50-year-old designs are described as modern. This inevitably leads to the conclusion that the MoD is implicitly defining modern equipment as any equipment that was procured in the last few years, rather than actually based on new designs. (Gorenburg 2012)

As evidence of the State Armament Programme's implementation until 2015 showed, the Russian defence industry still had not made the leap from 'dumb-iron' equipment to twenty-first-century sophistication.

Russia's intervention in Syria is a case in point. As Ruslan Pukhov argued, the performance of the Russian air force in Syria was impressive if compared to the 2008 war with Georgia, where it lost seven aircraft in blue-on-blue incidents within a matter of four days. From an international perspective, however, its technological level now matches at best that of the US air force during the 1991 Gulf War a quarter of a century ago. Weaknesses have been visible in Syria, particularly in terms of the sophistication of precision-guided munitions, a shortage of targeting systems, deficiencies in aerial reconnaissance capabilities, and a lack of long-range UAVs and attack drones (Pukhov 2016). The loss of two aircraft in short succession in autumn 2016, not over the theatre of operations in Syria, but during attempts to land on Russia's only aircraft carrier, the *Admiral Kuznetsov*, demonstrated that technological problems persist. Russian Defence Minister Shoigu himself acknowledged that a series of shortcomings in military equipment came to light during the Syria operations, when 'a number of design and manufacturing flaws were identified' (*The Moscow Times* 2016).

The continuing shortcomings in defence-industrial capabilities present an obstacle to the longer-term modernization of Russian military power for two reasons. First, the procurement of legacy systems in larger numbers as such will do little to help the armed forces overcome some major problems that were evident in the past. Clearly, shortages of tanks, armoured vehicles, submarines or combat aircraft *per se* were not primarily to blame for operational failures in Chechnya and in Georgia in 2008. Instead, it was a lack of advanced technologies required to create modern C4ISR[3] capabilities that impeded intelligence gathering, target identification and communications. This led not least to an excess of combat deaths and civilian fatalities. As discussed in chapter 5, the availability of sophisticated high-tech equipment is not a panacea and far from always strategically and operationally decisive. However, it can act as a force multiplier that shortens conflicts in certain circumstances and contains the intensity of destructive force having to be used, thus reducing casualties on both sides. In a conventional, state-on-state warfare scenario today, it is certainly clear that the technologically superior side will have a distinct advantage over a less 'modern' opponent.

The second obstacle to long-term modernization is the Russian defence industry's inability to deliver consistently across all categories of weapons systems. This means that the country's global power projection capabilities remain limited in certain areas. This is true in particular for naval power. The Russian navy had received the largest share of funding from the State Armament Programme 2011–20. This was because it was understood that of all branches of service it was in greatest need of modernization. As mentioned above, throughout the 1990s and 2000s, the country had no real operational navy to speak of. As only a limited number of ballistic-missile submarines and small

---

3 Command, control, communication, computers, intelligence, surveillance and reconnaissance.

coastal vessels were kept fully operational, the navy's role was reduced to just nuclear deterrence and coastal protection, depriving it of any genuine maritime power projection capability (Renz and Thornton 2012: 49). Naval power was not required during the small wars dominating the Russian armed forces' activities during the 1990s and its relevance for low-intensity missions is generally low. However, the ability to project power on the seas is an intrinsic requirement for a global military power. In spite of making the modernization of the navy a priority, this is where the State Armament Programme encountered the most serious problems (Cooper 2016: 49).

Deliveries under the State Armament Programme 2011–20 were restricted primarily to updated submarines and coastal vessels. Thus, the country is still at some remove from achieving the ambition of a 'blue water' navy. Ongoing difficulties, moreover, with the delivery of new and upgraded strategic bombers and transport aircraft to the air force also pose a further constraint on Russian global power projection capabilities. The lack especially of transport aircraft in sufficient numbers reduces the logistics' capacity required to deploy and sustain troops in the event of extensive out-of-area operations (Connolly and Sendstad 2016).

When the State Armament Programme 2011–20 was first introduced, Russia sought to make up for the defence industry's lack of expertise in certain high-tech areas with Western imports. Sanctions imposed in the aftermath of the annexation of Crimea, however, have closed its access to these markets for the foreseeable future. Moreover, the defence industry's reliance on other CIS states is a serious issue. The breakdown of deliveries from Ukrainian defence companies seriously exacerbated problems in the production of various weapons systems (Cooper 2016). Measures to counter the impact of Western and Ukrainian sanctions by producing the required technologies domestically have already been implemented. Although 'import substitution' has become a buzz phrase in Russian political rhetoric, it is far from a quick-fix solution to ongoing limitations in defence-industrial capabilities. Import dependency is very high in many vital areas, including microelectronics, advanced production equipment and information technology. For example, up to 95 per cent of all electronic components on Russian ships are reportedly foreign-made (Connolly and Hanson 2016: 17). Developing high-tech industries able to replace these imports with domestic products will require decisive efforts towards the diversification of economic activity and a high level of investment. In spite of all the rhetoric on 'economic sovereignty' and import substitution, Moscow's already chronically low levels of economic investment have decreased further in recent years. And, having to prioritize areas of spending during the continuing economic recession has meant, as Richard Connolly and Philip Hanson found, that 'spending on the military and social welfare has proven more resilient than that on areas essential to Russia's plans for high-tech industrial development'. The latter has thus been subject to heavy cuts (2016: 19). Ironically, this means that any decision to protect or to increase military expenditure in this time of ongoing economic stagnation will impede the modernization of the Russian armed forces in the longer term.

Conclusion

For almost two decades after the end of the Cold War, Russia struggled to transform what was left of the former Soviet military into one fit for the twenty-first century. This had serious implications not only for Moscow's ability to use armed force efficiently in various small wars and insurgencies that erupted across the former Soviet region throughout the 1990s; it also cast doubts on the country's ability to defend itself against a multitude of potential geopolitical threats and contributed to the loss of its status as a great power in the eyes of the world. By the end of Yeltsin's time in office, the lack of systematic reforms had left the country with a military that was widely seen as 'impoverished, demoralized and largely ineffective' and 'woefully inadequate to address the country's security threats' (Golts and Putnam 2004: 121; Barany 2005: 33). The failure to reform the armed forces for so long was, though, not due to a lack of recognition that such reforms were necessary or desirable. Central elements of the 2008 modernization programme had already been discussed during previous, failed attempts at reform. Moreover, Moscow's ambitions to maintain a military that could project power on a global level date back to the early 1990s. It was a combination of political and financial factors that prevented these ambitions from becoming reality.

The situation changed when Putin came to power and made the military reform agenda a priority. Aided by a steadily improving economic situation, the armed forces have undergone a remarkable turnaround in less than a decade. Structural and organizational reforms announced in 2008 were pushed through with unprecedented determination and followed up with a costly programme of rearmament. The annexation of Crimea and the air campaign in Syria demonstrated that the military had made considerable advances – in terms of tactical art, operational skills and equipment – in overcoming some of the shortcomings that led to limited outcomes in previous interventions. The performance of its troops in Crimea and Syria not only did much to restore the country's pride in its military, but it also revolutionized its image internationally and led to fears of a militarily resurgent Russia.

In stark contrast to the predominantly negative views of the Russian military in the West throughout the 1990s and 2000s, when the country's armed forces were more or less written off as a serious global actor, there are now fears that the country's capabilities had been seriously underestimated. As this chapter argued, and as Tor Bukkvoll has also noted, 'there might be some truth in that, but now overestimation is the greater danger' (2016b). Since the announcement of the modernization programme in 2008, the Russian military has experienced a revival, but it is important to bear in mind that this process is still ongoing and is by no means complete. The limited nature in both scope and size of the Crimean and Syrian operations does not allow for the conclusion that modernization efforts over the past few years have equipped the military with new capabilities to the extent that it now poses a real threat to European and transatlantic security. Moscow is still a long way off achieving its ambition of creating a global military capacity on a par with the world's strongest powers. A range of complex and deep-rooted limitations continue to stand in the way. The deepening economic crisis since 2009 has meant that the affordability of the Kremlin's longer-term military ambitions is yet again in doubt. Russia is catching up with other global military actors, but the extent of its military revival should not be overstated.

The next chapter discusses an important aspect of Russian military power that is often overlooked. This is the role its 'other' armed forces, or force structures, have played in the country's defence and security since the collapse of the Soviet Union. As this chapter has shown, the reasons for military reforms and the decision to strengthen the country's conventional military power are complex and can by no means be reduced to the intention of preparing for offensive action. The following chapter underlines further that Russian military power has a multitude of roles and functions and is as much about internal order and regime stability as it is about external threats and global competition.

Chapter 3

Russia's 'other' armed forces: the force structures

As Dmitri Trenin (2001: 74) noted, in order to understand the Russian military 'we have to consider the whole complexity of the meaning of the military in Russia, where shoulder to shoulder with the "first army" – the armed forces – there is a second one'. This 'second army', which is the subject of this chapter, comprises a number of ministries and federal services that have under their command armed personnel and militarized formations. In Russia, all ministries and federal services with militarized formations and armed personnel, including the Ministry of Defence, are referred to as *silovye struktury* or *silovye vedomstva*. Literally, this translates into 'power-wielding' or 'force-wielding' structures and the term most commonly used to describe these in English-language analyses are the security apparatus, power ministries or force structures (Bacon 2000; Renz 2005; Taylor 2007). The force structures other than the regular armed forces are tasked predominantly, but not exclusively, with internal aspects of security. As discussed in chapter 1, ensuring domestic order and protecting regime stability has been a function of military power throughout the country's history and it has also been an important factor in the recent military revival. As such, an appreciation of the force structures' roles and functions is essential for our understanding of the role of the military as an instrument of Russian state power.

The force structures that have under their command militarized units are considered part of the military organization of the Russian Federation in accordance with the country's laws. As stated in the military doctrine issued in December 2014, the 'military organization of the State . . . is a complex of state administration and military command and control bodies, the Armed Forces of the Russian Federation, other troops, military units and bodies' (The Military Doctrine of the Russian Federation 2014: section 1, paragraph 8). The terms of service of uniformed personnel employed in these organizations in most cases is determined by the same laws applying to military personnel serving in the regular armed forces. Some of the force structures, such as the now defunct Federal Tax Police Service or Federal Service for the Control of the Drugs Trade (FSKN), are law-enforcement agencies, whose uniformed personnel are classified as law-enforcement personnel and not military personnel. All force structures employing military personnel are allocated conscripts with the exception of the Federal Border Guard Service. Following its incorporation into the Federal Security Service (FSB) in 2004, this was reformed from a heavily militarized organization into a structure manned entirely by professional personnel over the past decade (Nikolsky 2013a).

Russia's 'second army' is sizeable. According to *The Military Balance*, today almost 500,000 people are employed in what this publication terms 'paramilitary' forces as opposed to the around 800,000 active soldiers in the regular military (*The Military Balance* 2016: 485). These numbers in themselves indicate the ongoing importance attached to the preservation of internal security and stability. Some of the force structures have also been used in international settings in recent years, making up for some shortcomings of the regular armed forces in dealing with military operations other than war. Democratic states do not, as a rule, maintain an equivalent range of such quasi-military organizations. As a result, they have often been skimmed over in Western analyses and as such remain poorly understood, because they simply do not fit into existing conceptual frameworks. This chapter shows that, as argued by Trenin, the role of Russia's military cannot be understood without consideration of the force structures. The multitude of roles that they fulfil, as well as their sheer combined size, means that their importance, potential influence and political impact is at least as significant as that of the regular armed forces, and they are central to the achievement of Russian policy objectives, both domestically and internationally, in various ways. Starting with an outline of the force structures' origins, the chapter discusses their functions and role as an instrument of state power and the various attempts at reforming them throughout the post-Cold War years. It concludes with an analysis of the factors determining continuity and change in the sector.

The Russian force structures: background and origin

During Soviet times the number of state institutions maintaining armed personnel and militarized formations was limited to three: the Ministry of Defence, the Ministry of Internal Affairs (MVD) and the KGB. When the Soviet Union collapsed, these vast ministries were broken up into numerous separate entities. Entirely new structures were also created. One factor that influenced reforms of the Russian security apparatus in the post-Cold War years was the country's obvious need to adapt to new challenges posed by the new international security environment. The collapse of the Soviet Union presented Russia with a number of challenges that its existing security apparatus was not equipped to deal with. As discussed in more detail in chapter 4, not unlike post-Cold War changes in the security priorities of Western states, 'new' security challenges, such as drug crime and trafficking, terrorism and the proliferation of ethnic conflicts, moved to the forefront of security concerns facing the Russian leadership (Bacon 2000: 3; Renz 2007). Reforms of the Russian security apparatus therefore reflected the country's need to adapt to the new security environment, especially as the regular armed forces were ill-suited to deal with contingencies at the lower end of the conflict spectrum.

As this chapter will show, political motives for creating and maintaining a large number of powerful quasi-military organizations have been another major determining factor for reforms, or lack thereof, of the sector throughout the post-Soviet years. An important reason for the rise in their number under Yeltsin was a 'divide-and-rule' strategy, especially in the early post-Soviet years. Splitting the force structures into smaller institutions diffused challenges from potentially powerful and conservative elements in the former KGB and Ministry of Defence and strengthened the president's grip on power (Desmond 1995; Moran 2002). When Vladimir Putin rose to prominence in 1999, the political role and influence of the Russian force structures, and of the FSB in particular, became a more widely discussed issue. As is well known, Putin himself has a career background in the KGB and was director of the FSB from July 1998 until March 1999. When he was elected president in 2000 the growing number of *siloviki* – a term traditionally used in Russian jargon to describe employees in any of the force structures – in official posts caught the attention of analysts both in Russia and in the West (Kryshtanovskaya and White 2003; Renz 2006). As the chapter will show, however, the political significance of the force structures goes far beyond the appointment of individuals to political posts.

On the one hand, the chapter will demonstrate that the characterization of the force structures as a 'presidential bloc', maintained by the civilian elite for holding power and securing the regime, holds true today (Vendil Pallin 2007: 3). On the other hand, the perception of a clear break or revival of the force structures under the former KGB officer Vladimir Putin requires some contextualization. It is beyond doubt that Putin has maintained these organizations as an important source of state power and has bolstered the strength of individual force structures, such as the FSB. At the same time it is clear that there is a large degree of continuity from the Yeltsin years. As Dennis Desmond noted in 1995, expectations that Yeltsin would continue the path of democracy and reduce the role of the all-pervasive force structures did not come to pass. He had 'immediately moved to consolidate his power base and instead strengthen the security services' (1995: 134). As such, the political system enabling Putin to bolster the force structures and to rely on them as an important power resource had already been put in place by his predecessor.

An overview of the Russian force structures

The Interior Ministry (MVD)

The MVD is the institutional successor of the Soviet MVD. Until the announcement of significant reforms that removed the domestic troops from this ministry and turned them into the Federal National Guard Service (FSNG) subordinated directly to the president in 2016, the MVD was a hybrid organization serving both as a law-enforcement agency and a military service. Until the 2016 reforms it was also the second-largest Russian force structure (after the Ministry of Defence) with more than one million employees. The range of tasks fulfilled by its various departments was very broad. Its law-enforcement element continues to be in charge of crime fighting and other 'traditional' police assignments, including road traffic safety. Due to endemic corruption amongst its personnel, the Russian police have enjoyed a low level of public trust throughout much of the post-Soviet period. In order to deal with this situation, reforms of the police service and a complete rebranding and renaming from *militsia* to *politsia* was announced by then-President Medvedev in 2010.

Until 2016 the MVD's militarized element consisted of approximately 200,000 internal (or domestic) troops, a legacy of the Soviet period. Their tasks included supporting the police in maintaining public security and contributing to the defence of Russian territory and its borders. MVD troops were heavily involved in both Chechen campaigns, where their conduct and inefficiency was strongly criticized, including by commanders of the MoD and FSB troops that were also involved in the conflict and sought to shift blame for military failures to the MVD. Although the domestic troops received combat training and, in the words of Gordon Bennett were 'made to look like a small army', they were not equipped or trained for fighting a low-intensity conflict like Chechnya (2000a: 15). As was the case with the regular armed forces discussed in the previous chapter, the need for reforming the interior troops had been an ongoing discussion since the 1990s. However, significant plans for transformation, such as the idea of turning them into a separate service similar to the US National Guard, did not come to fruition until 2016.

According to Brian Taylor, the MVD was the 'neglected stepchild of the Soviet power ministries', whose influence was always weaker than that of the Ministry of Defence and the KGB (2007: 7). Although the MVD did not experience serious institutional changes and was not split into numerous entities like the KGB up until 2016, this did not mean that the ministry was able to consolidate or strengthen its influence throughout the post-Soviet period (Taylor 2011: 46). On the contrary, in comparison to the institutional strengthening of the FSB throughout the Putin years and the attention and finances expended on reforming the regular armed forces since 2008, the MVD continued to take a relatively marginal position in the Russian security apparatus. The lack of attention paid to substantially reforming the MVD throughout much of the post-Soviet period came to an abrupt end in April 2016. With next to no prior discussion, the decision to create a new Federal Service, the National Guards, subordinated directly to the president, was announced. The new service was to be based on the assets and personnel of the internal troops as well as on the ministry's highly trained special assignment units, such as the riot police OMON and the rapid response unit SOBR. As part of the same round of reforms, the Federal Migration Service and drug enforcement agency, the FSKN, which is discussed in more detail below, were disbanded as separate services and subordinated to the MVD. In spite of these additions, the removal of the interior troops and special assignment units have significantly decreased the size and scope of the MVD and as such weakened its position vis-à-vis other powerful force structures. At the same time, the reforms meant that with the removal of its distinctly militarized element, it has become a more 'modern' institution at least in the sense that it now corresponds more closely to the remit of interior ministries in Western states.

The Ministry for Civil Defence and Emergency Situations (MChS)

The MChS emerged from the Russian Rescue Corps, which had been set up by RSFSR President Boris Yeltsin in response to the need for an integrated system of rescue and emergency response, in 1990. It was designated a State Committee by presidential degree in 1991 and in the same year more than 20,000 military personnel serving as civil defence troops under the Ministry of Defence were transferred to this new structure. With the integration of the around 250,000-strong State Fire Service in 2002, it has since become the country's third-largest force structure. The MChS and its predecessors were created in response to the need to deal more effectively with the consequences of natural and man-made disasters. Memories of the Soviet force structures' failure to cope with catastrophes, such as the Chernobyl nuclear reactor incident in 1986 and the earthquake in Armenia in 1988, informed this decision (Thomas 1995: 227). The ministry's tasks are extremely varied and the specialization of its personnel is wide-ranging. The MChS's largest element, the State Fire Service, is responsible for the work of fire services throughout the country. Its rescue forces provide rescue services and respond to the nationwide emergency telephone service. The civil defence troops, which underwent several transformation attempts, were reorganized into Military Rescue Units by presidential decree in 2011. The units comprise approximately 24,000 employees, of which around one-third are military personnel, both conscripts and professionals (MChS website: 'Spasatel'nye voinskie formirovaniia'). The aim of this reorganization was to make these troops more efficient by turning them into permanent readiness units. According to the MChS website, the change resulted in considerable improvements of the units' mobility and preparedness for responding to crisis situations (MChS website: 'O grazhdanskoi oborone').

The MChS has been compared to the US Federal Emergency Management Agency (FEMA) and both organizations are fulfilling a number of corresponding tasks. However, FEMA is a civilian organization and, with its around 14,000 personnel, considerably smaller than its Russian counterpart (FEMA website). As a quasi-military organization, the MChS does not really have a direct equivalent in the West to which it can be compared. The decision to keep a military element in this ministry as well as in a range of other force structures can be explained at least partially by the political leadership's need to avoid exacerbating the problem of creating vast numbers of unemployed former service personnel after the 'downsizing' of the armed forces at the end of the Cold War. Moreover, the military experience of civil defence troops was also likely to be useful in the traumatic environment of disaster situations. During the Soviet era, specialized civil defence subunits were maintained in order to provide assistance to the population in the event of bombing raids and nuclear, biological or chemical attacks (Aleinik 2006). Throughout the post-Soviet period the civil defence troops and Military Rescue Units retained such a wartime role by law and continue to be tasked with the organization and coordination of Russian military forces for the purpose of civil defence not only during natural and man-made disasters, but also during wartime.

Right from the start of the ministry's existence, its civil defence troops were able to gain considerable experience in providing crisis response and assisting civilian populations caught up in armed conflicts, both in Russia and internationally. In the aftermath of the collapse of the Soviet Union and as discussed in detail in chapter 4, ethnic conflicts erupted on the former Soviet territory and within Russia's own borders. MChS troops were involved in aiding civilian populations in Tajikistan, Transnistria, North and South Ossetia as well as Abkhazia during the 1990s. Civil defence soldiers also operated alongside regular armed forces, MVD and FSB troops during both Chechen conflicts, where they restored vital services in the Chechen capital Grozny, provided shelter, food, water and medical aid and engaged in de-mining activities for humanitarian purposes (Grau and Thomas 1999).

In the aftermath of Russia's five-day war with Georgia in August 2008, MChS troops provided aid to civilians in South Ossetia and to refugees that had fled the war zone and diffused mines and other explosives in civilian areas. The MChS also soon contributed to international crisis response and humanitarian operations. As such, it was a central component in a nascent Russian multilateralism which, as discussed in chapter 1, is an important function of the country's military power. The ministry's international activities are formalized in forty intergovernmental agreements and fifteen international statutes, detailing its partnerships with a range of international organizations, including the UN, NATO, the European Union and the Commonwealth of Independent States. A major area of MChS's international activity since 1993 has been in the sphere of humanitarian operations under the aegis of the UN High Commissioner for Refugees (UNHCR) and the World Food Programme (WFP). In cooperation with these UN agencies, MChS troops have built and equipped refugee camps, airlifted aid and participated in humanitarian de-mining efforts during conflicts in the former Yugoslavia, Central Africa and Lebanon. Until all practical cooperation was stopped in the aftermath of the Crimea crisis in spring 2014, the MChS had also sought close ties with NATO in the sphere of civil emergency response. The MChS's cooperation with NATO focused on emergency preparedness, civil-military cooperation and disaster management. Following a Russian proposal, the Euro-Atlantic Disaster Response Coordination Centre was established at NATO headquarters within the framework of the Partnership for Peace programme in 1998 (NATO 2006). In addition to large-scale exercises, which have previously involved both MChS and NATO forces, the Centre coordinates requests and offers of assistance from

NATO members and partner states. It has been involved in numerous operations involving the coordination of relief supplies to refugees, and providing assistance to civilian populations following natural disasters, such as forest fires, floods and earthquakes.

What distinguishes the MChS from many other force structures is that the ministry and the activities of its troops have attracted very little criticism throughout the post-Soviet era, both in Russia and abroad. This has mostly to do with the specificity of the tasks assigned to the ministry, which are to provide aid to civilians and organize rescues and have therefore been seen as 'laudable and uncontroversial' (Galeotti 2002: 50). In comparison to the failures, especially of the regular Russian armed forces involved in military conflicts throughout much of the post-Soviet era, the civil defence troops have also distinguished themselves with their efficiency. During the Chechen campaigns, for example, MChS troops were seen to be the most organized and effective federal force (Stepanova 2005: 142). Internationally, the MChS's focus on humanitarian operations has made its participation in cooperative efforts relatively straightforward. As discussed in more detail in chapter 4, the 'peacekeeping activities' of regular Russian armed forces in other former Soviet states throughout the 1990s were criticized in the West for being heavy handed and contravening generally accepted standards. The deployment of MChS troops had less potential for causing political tension than peacekeeping or stabilization tasks, because it did not include the application of military force.

The non-political nature of the MChS's international engagements has allowed the Russian leadership to engage even in those conflicts that it saw as too contentious for the involvement of regular armed forces. As such, it has been MChS troops above all other Russian military formations that have been used for cooperation in multilateral military cooperation. For example, in 2001 and 2002, the MChS provided humanitarian and medical aid to Afghanistan in cooperation with the UN and individual states (the United Kingdom, France and Germany) (Stepanova 2005: 145). In 2005, it delivered medical supplies to Iraq as part of a Russian–German operation under the auspices of the UN (Blinova 2005). The international image of the MChS as an uncontroversial provider of humanitarian aid was severely tainted when it was put in charge of sending 'humanitarian convoys' to eastern Ukraine in August 2014. These were widely suspected of being used as a cover for the delivery of Russian weapons to separatist rebels and accused of breaching international rules on the delivery of humanitarian aid (Roffey 2016: 45–7).

The Federal Security Service (FSB)

The FSB is the major institutional successor of the Soviet KGB and the largest and most influential of the Russian security services. As a federal service it is headed not by a minister, but by a director answerable directly to the president. Given the heavily influential role the KGB had played in the Soviet regime and the involvement of its leadership in the 1991 coup attempt, both Gorbachev and Yeltsin made the dismantling of the KGB a priority. Gorbachev abolished the KGB as an institution in 1991 (Bennett 2000b). It was split into a number of smaller units and, following several rounds of reorganization, the FSB was established by federal law in 1995. The FSB's size and sphere of responsibility grew significantly following Putin's election as president in 2000. As part of extensive reforms of the security sector in 2003, the FSB absorbed a significant portion of the assets and functions of the Federal Agency for Government Communication and Information (FAPSI), which had been established as one of the KGB's successor organizations in 1993. During the same round of reforms, the Border Guard Service, including the Border Guard Troops, was integrated into the FSB, increasing its numerical strength by approximately 160,000 personnel. During Soviet times the border guards operated under the auspices of the KGB and they had been transformed into a separate service by Yeltsin in 1993 (Bennett 2002). The reintegration of the Border Guard Service into the FSB significantly strengthened the latter's power and potential influence, leading to increasing concerns about the recreation of the KGB (Bacon and Renz 2003).

The FSB today is tasked with many of the functions carried out by its institutional predecessor and its growing power throughout the post-Soviet years demonstrates that internal control and regime stability continues to be a major factor in Russian security and defence policy. Reflecting the variety of structures under its command, the FSB's areas of responsibility are wide-ranging. They include intelligence and counterintelligence activities, dealing with organized crime and corruption on a national and international level, counter-terrorism, ensuring information security and the guarding of Russia's vast state borders. The FSB's economic security service is in charge of investigating economic crimes and monitoring financial activities, including those of foreign financial corporations with offices in Russia (Bacon et al. 2006: 157–8). The FSB is playing an important role in Russian cybersecurity and counterintelligence operations involving internet infrastructure. These are conducted by the FSB's Information Security Centre, which was created on the basis of FAPSI's Department of Computer and Information Security. The Centre works closely with the MVD's Directorate K in charge of cybercrime (Taia Global 2015).

The FSB can be described as a quasi-military organization as it employs both civilian and military personnel. Its most prominent militarized elements are the special assignment units (*spetsnaz*) in charge of counter-terrorism, *Alfa* and *Vympel*. Both were established during the Cold War. *Alfa* was created in 1974 within the KGB's 7th Directorate (Surveillance) in view of the upcoming Moscow Olympics. A counter-terrorism outfit similar to the German GSG-9, which had been created in response to the 1972 Munich Olympic terrorist attacks, was seen as essential in order to provide security at this event. *Vympel* was established in 1981 as part of the KGB's 1st Chief Directorate (overseas espionage) predominantly for missions abroad, such as the war in Afghanistan (Galeotti 2013: 35–43). Today these units operate as part of the FSB's Special Operations Centre, which was created in 1998 by then-director of the FSB Vladimir Putin. The rationale for creating this centre was to bring under one umbrella all of the service's special assignment units for a coordinated approach to the fight against international terrorism and extremism, as well as organized criminal groups (FSB website 2013). Both units were fighting alongside MVD and MoD troops in Chechnya and were centrally involved in the counter-terrorist operations ending the hostage crises in a Moscow theatre in 2002 and in Beslan in 2004 (Renz 2005). An amendment to the law 'On Counteracting Terrorism', which had been adopted in March 2006, granted the president the power to send FSB units to fight terrorism abroad (Remington 2009: 51; Soldatov and Borogan 2011: 249). At a meeting of the FSB collegium in February 2016, Putin openly acknowledged that FSB personnel were actively involved in the intervention in Syria, thanking their 'counterterrorism units active within the country' for their efforts (Putin 2016b).

Military personnel are also employed in the FSB's Border Guard Service. Given the vastness of the country's territory and the length of its borders, border guarding has been a historically important and difficult task for the Russian military, as porous borders that were hard to protect have been an important source of insecurity. During the Soviet era, the full length of the country's borders was protected by militarized border guard troops. Having been subordinated to the KGB in 1957, these troops were comparable in numerical strength and equipment to some countries' entire armed forces, and by the time of the collapse of the Soviet Union they had under their command around 190,000 personnel. During their existence as a federal service from 1993 until 2003, the Border Guard Troops continued to operate essentially as a military organization. Substantial reforms occurred only following their resubordination to the FSB in 2003. The principle of 'linear' protection of the entire length of Russia's border was abandoned in favour of relying on intelligence, reconnaissance and technology. The border service's personnel was reduced to around 100,000 personnel, all of whom are professionals. Its operating procedures remain militarized only at some particularly troublesome stretches of border (Nikolsky 2013a).

The Federal Service for the Control of the Drugs Trade (FSKN)

The FSKN was created by presidential decree in 2003. It was based on the personnel and material assets of the Federal Tax Police Service, which had been created by Yeltsin in 1992 and the disbandment of which was announced in the same decree. Although it was abolished as a separate service and subordinated to the MVD in 2016, this force structure deserves attention in this chapter, because the circumstances of its creation and dissolution exemplify many of the drivers and problems pertaining to reforms of the Russian security apparatus in the post-Soviet era as discussed further below. The FSKN is an interesting case because, unlike the majority of other force structures, it did not emerge from one of the Soviet power ministries. Instead, it was created from scratch and in response to a 'new' security challenge. Until the collapse of the Soviet Union, the scale and the trade and consumption of illegal drugs in Russia had been negligible compared to the threat this issue posed to states in the West (Schaffer Conroy 1990: 457). The opening of borders resulted in the rapid expansion of the drug trade and consumption. Over the next two decades Russia developed one of the world's largest populations of injecting drug users and also witnessed a fast growing HIV/AIDS epidemic (United Nations 2008).

Initially, the Russian leadership sought to counter these developments by using the combined efforts of the MVD, the FSB and the Border Guard Service. However, this never resulted in a coherent counternarcotics strategy and the expansion of the drug market into the country continued unabated (Paoli 2002). In order to enable a more effective approach to the fight against drugs, the FSKN was created with an establishment strength of 40,000 personnel. The FSKN was often compared to the US Drug Enforcement Administration (DEA). Although their basic tasks were indeed similar, there were also two important differences. First, with its establishment strength of 40,000, the FSKN was almost four times the size of the DEA with just around 10,000 employees. Second, whereas the work of the DEA is overseen by the US Department of Justice, the FSKN was answerable directly to the Russian president, raising questions of transparency and accountability. The FSKN was created in part in response to the need for better coordination of Russia's counternarcotics strategy. This meant its designated remit included stemming both the supply side and the demand side of the illegal market. From the outset, there were concerns that as a force structure the FSKN would emphasize law-enforcement operations at the expense of counter-demand interventions, such as educational work and dealing with social problems. Throughout its existence there were frequent critical reports about its repressive approaches, which tainted its image as an institution able to pursue a balanced counternarcotics strategy (Renz 2011: 64–6).

Around half of the FSKN's staff was classified as law-enforcement personnel. This included a number of special assignment units subordinated to the FSKN Department for Special Operations and Protection's 5th Operations and Combat Division. The major tasks of these units was to provide armed support during operations against organized drug crime as well as to offer protection to senior FSKN officers (Nikolsky 2014). These units have also been active in major conflict situations both inside Russia and internationally. In the North Caucasus and in Chechnya they fought alongside military personnel to stop the trading of drugs by insurgent groups. They also participated in joint operations with US military personnel and Afghan counternarcotics forces to raid and close down drug laboratories near the Pakistani border (Renz 2011: 60, 69). Like the MChS, the FSKN was tasked to deal with what are often referred to as 'new' security challenges, which transcend national boundaries and whose solutions require multilateral cooperation rather than rivalry between nation states. This understanding was one of the factors leading to the creation of the FSKN in 2004 and its statute emphasized cooperation and exchange of information with international organizations and foreign partners as one of its central tasks (Presidential Decree 976, 2004).

As the vast majority of opiates enter Russia from Afghanistan via the Central Asian states, multilateral cooperation with the former Soviet states in Central Asia, also within the framework of the Collective Security Treaty Organization, was the central focus of the FSKN's international cooperation. However, it also signed agreements with many other states, including the US, China, neighbouring Nordic countries and a number of countries in South-East Asia and Latin America (Renz 2011: 66–7). Especially in the early years of the service's existence, FSKN officers visited foreign counterparts, for example in the US, the

UK and in Sweden, to familiarize themselves with their experience in counter-supply and counter-demand work. As part of a NATO–Russia Council project for counternarcotics training in cooperation with the UN Office for Drugs and Crime, FSKN instructors contributed to the training of drug-enforcement officers in Central Asia, including in Afghanistan (Renz 2011: 68). Throughout its existence, the service was embroiled in a number of scandals and corruption charges and its effectiveness in stemming the trade and abuse of drugs in the country was generally evaluated as low. However, its dissolution in 2016 came as a surprise to many as it had been assumed that, as the main agency in charge of international cooperation in the fight against drugs, such a step could only be counterproductive (Nikolsky 2013b). The FSKN's successor will continue to exist as a separate service, albeit under the umbrella of the MVD. The extent to which Russia's engagement in multilateral counternarcotics cooperation will develop under the new set-up is a question for the future.

The Federal National Guard Service of the Russian Federation (FSNG)

The FSNG is Russia's youngest force structure and was established in April 2016. It was created on the basis of the MVD interior troops, the MVD's riot police OMON and rapid reaction units SOBR, MVD security personnel guarding sensitive government and corporate facilities, and the Okhrana company, which provides guard services to private customers (Nikolsky 2016; Presidential Decree 157, 2016). The new service is a sizeable outfit and with the combined numerical strength of its component parts, it has been estimated to comprise between 320,000 and 430,000 personnel with the 170,000–200,000-strong former interior troops making up its largest element (Kramnik and Bogdanov 2016; Nikolsky 2016). Soldiers that previously served in the interior troops retained their military rank and the service's director announced shortly after its creation that the approximately 30,000 OMON and SOBR troops would transfer from law enforcement to the status of military personnel by 2018 (I. Petrov 2016).

According to the presidential decree announcing the creation of the FSNG, the new service's remit is to ensure the security of the state and society as well as to protect human rights and the freedom of citizens. Further, its tasks include cooperation with the Interior Ministry in protecting public order during emergency situations and contributing to the fight against terrorism and extremism, territorial defence and border protection, the guarding of important government facilities and overseeing the licensing of private security firms and control of civilian firearms (Presidential Decree 157, 2016; Federal law 226-FZ, 2016). According to Putin, the FSNG will continue to fulfil the same tasks that were previously assigned to the interior troops, OMON and SOBR forces under the auspices of the MVD (Putin 2016a). As the Russian security experts Ilia Kramnik and Konstantin Bogdanov cautioned, however, it remains an open question whether new and additional tasks could potentially be assigned to the service in the future. In particular, they noted that if the FSNG was to be given investigative powers in support of its work in the sphere of terrorism and extremism, the country would gain not only a new force structure, but an entire new intelligence agency (2016).

Although the announcement of the FSNG in April 2016 came as a surprise at the time, the possibility of establishing a National Guard, not least because of the need to substantially reform the interior troops, had been discussed since the early 1990s (Renz 2012a: 211–12). Still, questions were raised about the timing of the decision to implement this idea in 2016. As Kramnik and Bogdanov argued, the existence of a

National Guard as part of a country's security sector as such is nothing unusual. In some ways, the US National Guard's dual function as both a military reserve and interior force for use in disaster management and public order maintenance during mass unrest is not wholly dissimilar from that of the FSNG. However, they also noted that National Guards are often a feature in non-democratic states, especially in situations when the incumbent regime is driven by actual or perceived threats to its authority. In their words, 'such structures are difficult to incorporate into the country's existing state structures, because they are set up as a guard against internal conspiracies or coups' (Kramnik and Bogdanov 2016).

When the creation of the FSNG was announced, the majority of analysts, including pro-Kremlin observers in Russia, agreed that the decision was taken not only to improve the efficiency of the interior troops that had to date been left largely unreformed, but also in order to ensure regime stability and to quell potential public disorder in view of the upcoming parliamentary and presidential elections (Nikolsky 2016). Russian experts writing in *The Moscow Times* expressed concerns that civil society would be the main target of the FSNG's activities. They suspected the new service had been set up because it was subordinated directly to the president and, having removed 'the unnecessary link – that of minister – between the commander-in-chief and the head of the National Guards', its troops could now be deployed with impunity (*The Moscow Times* 2016). As Mark Galeotti noted (2006), the National Guards 'have little actual role in fighting crime or terrorism' and the removal of OMON and SOBR from the investigative element of the MVD is likely to decrease their crime-fighting abilities, rather than strengthen them. Based on this observation, he concluded that the creation of the FSNG was above all the result of the Putin regime's serious fear of potential public unrest.

It appears that, not unlike the FSKN, the FSNG was created from scratch to deal with a 'new' security challenge. However, on this occasion this was not a transnational threat requiring international cooperation, such as drug crime or terrorism. Instead, it reflected the distinctly 'domestic' concern over perceived threats to internal order and regime stability. These issues had already been flagged up as a major worry in the military doctrine issued in 2014 as discussed in more detail in chapter 5, and as also pointed out by Kramnik and Bogdanov (2016). For the first time in post-Soviet history, this military doctrine referred to information activities in the section on 'domestic military dangers'. Emphasizing the perceived threat of 'information influence over the population . . . aimed at undermining spiritual and patriotic traditions', the 2014 doctrine demonstrated the leadership's anxiety over social stability and fear of outside interference in Russia's domestic affairs (Sinovets and Renz 2015: 2). As such, the creation of the FSNG supports the argument that concerns over internal security and regime stability are an important element in Russia's military revival.

Other force structures

In addition to the institutions outlined above, there are a number of other Russian ministries and services that are generally regarded as component parts of the country's security apparatus. The Federal Guard Service (FSO) is an offshoot of the Soviet KGB. Its major missions are to protect the president and other Russian high-ranking officials, as well as buildings and strategically important infrastructure. When the Federal Agency for Government Communication and Information (FAPSI) was disbanded in 2003, the bulk of its personnel and assets were transferred to the FSO. A new service within the FSO, the Service for Special Communications and Information (SSSI) was created and made responsible for the organization, development, maintenance and security of special communications of all state bodies conducting signals intelligence collection, analysis, and exploitation (similar to the functions of British GCHQ and the US National Security Agency) (Renz 2005: 575). The FSO's activities are often conducted in close cooperation with the FSB, the foreign intelligence service SVR and the MVD (Gamov 2000). The FSO has a clear military element as it has under its command a brigade and two regiments, including the prestigious Presidential Guards Regiment. With the exception of the latter, which is still staffed in part by conscripts, its personnel are serving on a professional basis. The overall numerical strength of the FSO is estimated at around 30,000 personnel. This includes around 8,000–9,000 military personnel on protection duty (Nikolsky 2013c). Although the FSO and its activities rarely attract public attention these days, this was not always the case. The service was highly visible and controversial during the first part of the 1990s, when the head of the Presidential Protection Service (SBP), which existed as a separate service from 1993 until 1996, was headed by Aleksandr Korzhakev, a close ally and confidante of Yeltsin (Renz 2005: 576).

The Foreign Intelligence Service (SVR) is another successor organization of the Soviet KGB. It is the direct successor of the KGB's First Chief Directorate. Its functions are similar to those of its foreign counterparts, for example the CIA or the British MI6. The service is fairly technically minded with intelligence gathering being its central role. The SVR primarily conducts human intelligence operations, though it also maintains a cyber warfare capability about which very little is known. It is also tasked with contributing to counter-terrorism activities and the fight against organized crime in cooperation with other Russian force structures (the FSB in particular) and with relevant foreign counterparts. Owing to the nature of the SVR's activities, information on the service's establishment strength is classified. This has previously been estimated at 10,000–15,000 personnel, including around 300–500 special assignment forces in charge of guarding Russian diplomatic missions and Russian officials visiting zones of conflict (Bennett 2000c; Mukhin 2000; Nikolsky 2014).

The Ministry of Justice (Miniust) is one of Russia's less obvious force structures. It is regarded as such, because it is responsible for the agency dealing with the administration of Russia's penal system, the Federal Service for the Execution of Sentences (FSIN). Miniust was given responsibility for this service in 1998 in response to a request by the Council of Europe requesting Russia to conform more to European prison rules as a condition of Russia's membership in the Council (Renz 2005: 569). Previously, the prison service had been run by the MVD. The FSIN is in charge of a large number of uniformed prison service personnel and also of militarized special assignment units subordinated to its regional branches. The task of their around 4,000 troops is to deal with riots at prisons, escaped prisoners and protecting FSIN officers on dangerous missions (Nikolsky 2014). They were also involved in the Chechen conflict, where they cooperated with the regular armed forces and MVD troops in the liberation of populated areas and the organization of checkpoints (Renz 2005: 570). Other institutions often included in the range of force structures because of the specificity of their tasks, the position they take in the system of executive power, or because they have under their command special assignment units or military servicemen, are the following: the State Courier Service, the Federal Customs Service, the Main Directorate for Special Programmes under the President, and the Presidential Directorate for Administrative Affairs (Renz 2005; Taylor 2007; Vendil Pallin 2007). Due to their small sizes and the specificity of their tasks, little is known about these structures and they have not been involved in public political controversies.[4]

Post-Soviet era change and transformation of the Russian force structures: drivers and problems

During the early post-Soviet years, a number of force structures were prioritized over the regular armed forces when it came to funding and reform efforts. As Brian Taylor noted, given the situation in Chechnya, the growing danger of terrorism and increasing levels of organized crime, 'this focus made sense' (2007: 12). The existence of well-maintained and effective force structures, such as, for example, the MChS and its civil defence troops, was essential during the 1990s, as these could make up for some of the regular armed forces' shortcomings. The existence of force structures specializing in emerging security challenges also arguably made it easier for Russia to establish multilateral and international security cooperation in certain areas. For example, Russian cooperation with NATO involving regular troops always remained embryonic not least due to diverging geopolitical interests and continued mutual mistrust regarding the other's priorities and intentions, as discussed in more detail in the next chapter. In contrast, as discussed above, MChS's integration with NATO's Civil Emergency Planning did not encounter such problems (Renz 2007). The involvement of Russian regular armed forces in efforts to tackle the opium business in Afghanistan would have been unthinkable. Within

---

4 All of the force structures discussed in this chapter have websites offering various degrees of information. These can be accessed via the Russian government internet portal, www.gov.ru. The official website of the National Guards of the Russian Federation can be found at www.rosgvard.ru.

the framework of the FSKN, albeit on a very small scale, such cooperation was possible. These examples suggest that when cooperation is based on shared interests and not overshadowed by political concerns, multilateral military cooperation between Russia and Western states and institutions can be achieved.

Having said this, all is not well with the Russian force structures after more than two decades of transformation and restructuring. Concerns over their ongoing inefficiency, the militarization of the Russian state and society as a result of such a large number of militarized services, as well as over the potentially excessive influence and political role of these institutions, are certainly justified. The collapse of the Soviet Union left the Russian Federation with millions of citizens that had been employed in the country's three vast force structures, the MoD, the MVD and the KGB. An immediate and drastic reduction of this workforce would have exacerbated the problem of unemployment in a country that was already struggling to accommodate hundreds of thousands of soldiers withdrawn from other former Soviet states and Warsaw Pact countries. However, the maintenance of hundreds of thousands of uniformed personnel in various structures until this very day shows that their prioritization was not only due to socio-economic considerations. The Russian force structures have undergone several rounds of significant change over the past decades, but in many cases this resulted in little more than the reshuffling of tasks and personnel, rather than reforms aimed at streamlining and making the security apparatus more efficient and accountable. When the Federal Tax Police Service was disbanded in 2003, its assets and personnel were transferred to a new Federal Service, the FSKN. Similarly, FAPSI, which was abolished at the same time, was distributed between a number of already existing services and the FSB and FSO in particular. The MVD interior troops, which had long been seen as outmoded and inefficient, were reassigned wholesale to a new organization, the FSNG, in 2016.

The maintenance of a large number of force structures throughout the post-Soviet era, in spite of several rounds of reorganization, has meant that a crucial problem pertaining to the Russian security sector has not been dealt with. This is the lack of coordination and cooperation between the different services, whose functions are often overlapping. To cite but two examples, several Russian force structures share responsibility for the areas of counter-terrorism and counter-drug operations. According to Russian law, the FSB, MVD, FSO, SVR and Ministry of Defence all have a role to play in counter-terrorism (RUSI 2007). In 2016 the newly created FSNG was added to this list. Force structures with a stake in counternarcotics included the FSKN until its subordination to the MVD in 2016, the MVD, the FSB and its Border Guard Service, as well as the SVR (Babaeva 2004). Problems of coordination and cooperation between the relevant agencies have continued to hamper the realization of comprehensive and effective counter-terrorism and counter-drug strategies.

The ongoing inability to coordinate the effort of Russia's various force structures has been exacerbated by the country's historical lack of experience in joint operations and the management of inter-agency task forces. Joint and inter-ministerial operations both in conventional warfare and low-intensity conflict, as also discussed in chapter 2, were not trained for during the Soviet period. Post-Soviet Russian ambitions towards developing such capabilities, which are commonplace between Western military organizations and civilian agencies, only started to emerge in recent years during large-scale exercises and operations (Norberg 2015).

Lack of coordination and overlapping functions between different force structures have created inefficiencies, leading to 'vast amounts of duplication and waste', as Mark Kramer has put it (2005: 219). This caused significant failures and loss of life, especially in Russian counter-terrorism efforts over the past decades. During both Chechen campaigns the lack of coordination between the different agencies created significant problems (Kramer 2005: 217–19). During the Beslan school siege, poor inter-agency coordination, and in particular between the FSB and MVD, resulted in a lack of operational command and control. This allowed vigilante action to significantly influence the course of events, leading to the death of more than 300 people, including many children (Forster 2006). When it comes to counternarcotics, lack of inter-agency coordination caused the death of an MVD *spetsnaz* operative and the wounding of two FSKN officers during an operation against an organized crime group in Nizhnii Novgorod in 2006. Being unaware of each other's actions, the parties opened fire, mistaking each other for members of the criminal gang (Rushev 2006).

Being aware of the problems created by poor coordination between the various force structures, the Russian leadership has implemented numerous efforts to alleviate the situation. In addition to the training of large-scale, multi-agency operations during exercises from the second part of the 2000s onwards, efforts have included the creation of various supra-agency coordinating bodies. Following a number of significant terrorist events within the Russian Federation, including Beslan and the theatre hostage crisis in Moscow in 2002, new counter-terrorism legislation adopted in 2006 assigned the FSB as the main agency in charge of countering terrorism. Alongside the new law, a new National Counterterrorism Committee was established in the same year, which was tasked with both the coordination of all agencies involved in counter-terrorism activities and the shaping of policies and legislation in this sphere. Headed by the director of the FSB, membership in the committee includes, amongst others, the ministers of defence, interior, foreign affairs, emergency situations (MChS), justice, health, education and transport as well as the directors of the SVR, FSO, the Prosecutor General, top military leaders, the deputy heads of the presidential administration and Security Council and deputy speakers of both houses of parliament (Saradzhyan 2006: 177).

When it comes to counternarcotics, a new permanent coordinating body, the State Anti-Drug Committee, was created in 2007. The intention was to improve inter-agency coordination in this area, also in recognition that a balanced approach to the problem, which did not overemphasize the law-enforcement element, was needed. Under the chairmanship of the FSKN director, the committee's membership included the leaders of most force structures, as well as the heads of relevant civilian bodies, such as the ministers of health and social development, education and science, and economic development. The committee's tasks also included the control of regional anti-drugs commissions and the preparation and presentation of proposals for improving anti-drug policy to the president (Renz 2011: 61–2). The creation of such committees can be seen as a step in the right direction. However, as they coordinate activities at the federal level, their effectiveness in improving inter-agency cooperation on the ground is far from clear. Moreover, the creation of supra-agency bodies does nothing to deal with the issue of duplication and overlapping functions. In 2014, the National Defence Control Centre was established in Moscow. Enabling the coordination of efforts by all force structures, headed by the Ministry of Defence and including other economic, social and political authorities, the Centre was created to facilitate a 'whole government approach' to a large range of security challenges (Giles 2016: 25–6). Whether this Centre will lead to substantive improvements in coordination between the various forces involved in crisis situations is a question for the future (Tsymbal and Zatsepin 2015).

A significant factor accounting for ongoing problems in cooperation between the force structures is inter-service rivalry. The process
of reshuffling, merging, disbanding and recreating various force
structures throughout the post-Soviet era only exacerbated this problem. The abovementioned duplication and overlap of tasks across
several agencies inevitably leads to competition, because the directors of such agencies fear the encroachment of other structures into
their sphere of authority. Matters of resource allocation and lobbying
for an appropriate share of the budget have been an important bone
of contention between the leadership of different force structures
(Kalyev 2002; Kramer 2002; Galeotti 2013: 6). The wholesale dissolution of some powerful force structures, such as FAPSI, the Border
Guard Service, the Federal Tax Police Service and most recently
the FSKN, reinforced the need for lobbying by the force structures'
leaderships on behalf of their institutions and for currying favour
with the political leadership. Such a dynamic inevitably has come at
the expense of prioritizing true reform and efficiency in encouraging
rivalry, rather than cooperation.

During Yeltsin's presidency the strengthening of some force structures at the expense of the regular armed forces led to public displays
of conflict and jealousy between the Ministry of Defence, the MVD and
the security services (Moran 2002). Throughout the post-Soviet period,
various force structures were involved in highly publicized displays
of inter-service rivalry. One famous instance embroiling the MChS
in 2003 was dubbed by the Russian media the 'werewolves in uni-
form' affair. The then-head of the MChS security directorate, Vladimir
Ganeyev, was arrested alongside six police officers of the Moscow
Criminal Investigations Department, having been accused of extorting
money from local businesses. Ganeyev received a long prison sentence
and there was much speculation in the media that the scandal was

caused by ongoing tensions between Sergei Shoigu, the emergencies minister, and Boris Gryzlov, the interior minister, whose internal security directorate had been responsible for the arrests (Dobrolyubov 2013). Following the arrest of senior drug police officers on corruption charges in 2007, then-FSKN director Viktor Cherkesov published an open letter in the Russian media warning about the dangerous consequences of what he saw as the evolving 'war' of the force structures fanned by the ambitions of the intelligence services. His letter was widely seen as a thinly veiled attack on the FSB, which he accused of exceeding its authority in the selfish interest of its leadership without consideration of the consequences this would have for the country's national security (Cherkesov 2007).

When the FSNG was created in 2016, numerous analysts suspected that this was likely to create more tension between the force structures although the decision had been taken, at least in part, as a way of dealing with existing rivalries. Nikolay Petrov, for example, interpreted the creation of the FSNG and concurrent dissolution of the FSKN as a 'corrective to the power imbalance in the *siloviki*, where, in recent years, the Federal Security Service (FSB) and the Army have tremendously increased in power'. He also noted that one reason for the reform was to 'keep the *siloviki* busy and uncertain about their future' (N. Petrov 2016). Aleksei Nikolsky speculated that personal factors played a role in the reforms. In the run-up to the creation of the FSNG, longstanding rumours suggested that two close Putin allies, Viktor Zolotov, then-commander of the interior troops, and Viktor Ivanov, the director of the FSKN, were both vying for the position of interior minister. According to Nikolsky (2016), it was not entirely unlikely that Putin abolished the ineffective FSKN and created the FSNG with Zolotov as its director 'simply to resolve the delicate issue of his allies competing for various jobs'.

Concerns pertaining to the political influence of the force structures remain. Although they fulfil a number of important tasks in the sphere of Russian foreign and security policy, it is also clear that these vast institutions have consistently been used by the political leadership to strengthen and uphold the regime in power by installing personal allies in influential posts. This confirms further that the significance of Russian military power and reforms is complex and can only be understood if domestic political processes are taken into account. Many instances of reshuffling the security apparatus, as discussed above, were clearly instigated at least in part for political reasons, rather than by the desire to make the force structures as a whole more workable and efficient. The fortunes of individual force structures continue to be less the result of their competence as an institution, and more of their perceived loyalty and the closeness of their leadership to the political elite. This tendency goes back to the early Yeltsin years, when the three Soviet power ministries were fragmented in order to diffuse the potential power of any one of them and to create force structures explicitly loyal to the political leadership.

An extreme example of the importance of political loyalty in the force structure's fortunes is the Presidential Security Service (SBP), which was later subsumed into the FSO. The agency was considered highly politicized and had the right to conduct secret investigations when government officials were suspected of wrongdoing. According to Nikolsky (2013c), Yeltsin at the time even called the service 'the president's own KGB'. Aleksandr Korzhakov, who was involved in the personal protection of Yeltsin from 1985 until 1996, was believed to have interfered in executive decision making on several occasions. For example, the initial hesitation of Defence Minister Pavel Grachev to provide military support to Yeltsin during the October 1993 constitutional crisis was alleged to have been overcome thanks to the persuasiveness of Korzhakov. In the run-up to the presidential elections in the spring of 1996, Korzhakev controversially called for the postponement of the vote. He was dismissed from the SBP in 1996, and the service was subsumed into the FCO (Renz 2005: 577–8).

The establishment and rapid growth in size of the MChS under Yeltsin was also believed to have occurred for political reasons. Sergei Shoigu headed the MChS for 22 years from its very early stages as the Russian Rescue Corps in 1990 until his appointment as defence minister in 2012. He was explicitly loyal to Yeltsin during the 1991 August coup attempt and during the constitutional crisis in 1993. Shoigu's loyalty was rewarded in 1994 when his agency was turned into a ministry. According to Nikolay Dobrolyubov (2013), Yeltsin, who never fully trusted the regular armed forces, MVD troops and KGB successor organizations, had created the MChS as a powerful new force structure he could fully count on. As discussed above, the new FSNG created by Putin in 2016 was evaluated mainly as an instrument that would ensure the stability of the regime in power. Some analysts suspected the FSNG was established as a counterbalance not only to the FSB, but also to the regular armed forces which, as a result of successful modernization since 2008, had become increasingly powerful (*The Moscow Times* 2016).

Conclusion

The Russian force structures constitute an important element of Russian military power and component in the country's security and defence policy. As such, their roles and functions cannot be excluded from the study of military developments in the country. The large number of distinct entities in the Russian security sector is a legacy of the early post-Soviet years, when Yeltsin split the existing three Soviet power ministries into several smaller entities. Expectations that Yeltsin would continue on the democratic path and decrease the size and influence of the force structures over time did not come to pass. Although the respective influence of individual force structures has waxed and waned in line with changing security priorities and political constellations, they have been a consistently formidable resource at the disposal of the civilian leadership. In this sense, there is a significant element of continuity putting into context the idea of a sudden military revival, including that of the force structures and the FSB, under the leadership of Putin.

There are several reasons why the number of force structures was increased in the early 1990s and maintained until today. One such reason was the need to respond to 'new' security challenges that emerged after the collapse of the Soviet Union and the end of the Cold War and that the existing security structures were ill-equipped to deal with. These included the proliferation of ethnic conflicts in some regions of the former Soviet Union as well as within Russia's own borders, humanitarian crises, organized crime, terrorism and extremism. Especially in the early post-Soviet years, the need to build capabilities required for dealing with small-scale contingencies, rather than with 'traditional' military threats, appeared the most urgent. As a result, building the capacities of relevant force structures was prioritized over reforms of the regular armed forces. As the examples of the MChS's civil defence troops, the FSB's counter-terrorism elements and the FSKN's special assignment units show, the force structures were used to make up for the shortcomings of the regular armed forces in various low-intensity missions. Force structures dedicated to specific missions, and the MChS civil defence troops in particular, have also given Russia the opportunity to gain experience in multilateral military cooperation in a politically uncontroversial way. It is likely that if the activities of the NATO–Russia Council, which were suspended in the aftermath of the annexation of Crimea in 2014, should ever be resumed, any future military-to-military cooperation will involve force structures such as the MChS, rather than the regular armed forces.

Domestic political considerations have been another important reason for keeping the number of force structures high throughout the post-Soviet era. Increasing the number of ministries and services tasked with the provision of security and defence has allowed the civilian leadership to divide and rule and to prevent any one institution from becoming too influential and posing a potential threat to the regime. This was particularly important for Yeltsin, as he was wary of the role the KGB had played in the August 1991 coup attempt. Powerful and loyal force structures subordinated directly to the president have been maintained also to ensure public order and stability, an important prerequisite for bolstering and upholding Russia's great power status.

In Yeltsin's eyes, the major threat to his nascent regime came from revisionist elements in the political and military elites that sought to put a halt to the political and economic changes that had occurred since the fall of the Soviet Union. His strategy of divide-and-rule paid off when, during the 1993 constitutional crisis, military and security forces loyal to the president resolved the conflict in the executive's favour. Ensuring regime stability and public order has also been an important driver of force structure reforms under Putin. In contrast to Yeltsin, the major threat to regime stability in the eyes of Putin is not a revisionist political opposition but the potential for major public unrest caused by external influence over the population. This concern was formalized in the 2014 military doctrine as a domestic military danger for the first time, as discussed in more detail in the following chapters. However, it is a continuation of fears over outside and specifically Western 'information influence' as a threat to Russian national security that date back at least to 2000 (Bacon et al. 2006: 89–90). Increasing anxiety over the potential for public unrest caused by external interference – closely linked to concerns over the country's domestic sovereignty – explains the strengthening under Putin of those force structures able to deal specifically with domestic dissent and opposition, including the FSB and the newly created FSNG.

Developments in the force structures since the end of the Cold War provide further context for a better understanding of Russian military power and the reasons for its recent revival. The annexation of Crimea and subsequent actions have led to concerns in the West that efforts to bolster the country's military signified the preparation of further offensive action against neighbouring states and beyond. Like the previous two chapters, this chapter suggested that Russia's desire to maintain and strengthen its military power is about more than fighting wars and defeating opponents. Like the regular armed forces, the force structures have been used to deal with a wide array of tasks related to the country's security and defence, both domestically and internationally. These have included the need to deal with 'new' security challenges, such as international terrorism and organized crime, the protection of borders, and the involvement in multilateral military cooperation. The chapter also confirmed that Russia's military revival is as much about concerns related to domestic order and regime stability as it is about global power projection and competition. This is an important consideration when the implications of and possible reactions to the military revival are assessed.

Chapter 4

Russian uses of military power since 1991

With the annexation of Crimea in spring 2014, Russia forcefully expanded its territory for the first time since the creation of the Russian Federation in 1991. The Kremlin's willingness to use military force to conquer territory in blatant violation of international law led to understandable fears about what this apparent turnaround in approach might signify for the security of neighbouring states and of Europe at large. International concerns were heightened further when Russia intervened in the civil war in Syria in autumn 2015. Up until this point, the country had never engaged in unilateral military action beyond the borders of the former Soviet Union. Both operations seemed to suggest that an important watershed in the Kremlin's regional and global ambitions had been reached.

Russian actions in Crimea and Syria have been widely portrayed as a dramatic turn in Moscow's foreign policy. Some observers, often with reference to Putin's now infamous quote in which he called the collapse of the Soviet Union 'the greatest geopolitical catastrophe of the century', feared that the annexation of Crimea was only the start of efforts to materially recreate the Soviet Union (Van Herpen 2015: 3–5; Hall 2016). In the eyes of former US Secretary for Defence Leon Panetta, Russia's intentions in the region were unambiguously clear: 'let's not kid anybody: Putin's main interest is to try to restore the old Soviet Union. I mean, that's what drives him' (quoted in CSIS 2016).

Although the Russian Federation has never launched a military offensive against a Western state, the annexation of Crimea and operations in Syria were also interpreted as a threat to the West. There is the feeling today of a new quality of hazardous confrontation and assertions of a new Cold War have abounded (Kalb 2015; Kroenig 2015; Legvold 2016). In the eyes of some observers, Russia's involvement in Syria has set the world on a 'dangerous collision course', raising the risk of escalating tensions with the US that could lead to a Third World War (Dejevsky 2017; Kahl 2017). As discussed in the book's introduction, the perception of a 'revanchist' Russia threatening international security has already led to adjustments in the defence postures of many European states. At the extreme end of reactions, some commentators, such as the former NATO deputy commander and British General Alexander Richard Shirreff, have gone as far as to suggest that Russia was on course for war with the West. In Shirreff's words, 'under President Putin, Russia has charted a dangerous course that, if it is allowed to continue, may lead inexorably to a clash with NATO' (MacAskill 2016).

Various experts of Russian foreign and defence policy have since pointed out that the focus on territorial expansion and Moscow's intention to confront the West reflects a limited understanding of the operations in Ukraine and Syria. These actions did not occur in a vacuum. In order to explain the use of force in both cases, a combination of many factors, including status concerns, strategic interests, insecurity, historical ties and domestic developments, all variously need to be taken into account (see, e.g., Charap 2013; Allison 2014; Charap and Darden 2014; Averre and Davies 2015). Although these authors differed in their estimations of the degree to which Russian policies are being driven by 'revanchist' intentions, they all made the important point, implicitly or explicitly, that not only do reductionist interpretations of Russian actions not reflect the complexity of the situation, but the acceptance of such interpretations as fact is problematic in practice. As Roy Allison put it in the aftermath of the annexation of Crimea, 'practitioners as well as scholars need to draw on an explanatory framework that includes, but goes beyond, reliance on geopolitical categories and structural power, otherwise the current spiral of antagonisms between western and Russian leaders could intensify and have long-term profound consequences' (2014: 1255).

This chapter starts from the assumption that the motivations behind Russian military actions in Ukraine and Syria require detailed enquiry and cannot be reduced to vague notions like 'revanchism' or the intent to confront the West. As this book has argued so far, the reasons for and implications of Russia's military revival are not straightforward. Recent events can only be understood in the context of relevant developments in the country's history and politics, both domestic and international. This chapter discusses the Crimea and Syria operations within the framework of Russian uses of military force since the early post-Soviet years. The Kremlin has regularly used military power in pursuit of a variety of policy objectives since the early 1990s. Indeed, Russian military involvement outside of its own borders was significant from the outset. Although official numbers are not available, Pavel Baev estimated that by the end of 1992 around 27,500 Russian troops were engaged in various trouble spots in the region, increasing to around 36,000 by the end of 1993. His estimate for 1994 was as high as 42,000 troops, which also included those soldiers engaged in UNPROFOR in Bosnia discussed further below (1998: 218). As such, the country's preparedness to use military power as a tool of foreign policy is not a new development.

In order to understand what is new and what merely appears to be new about Russian motivations for using military force today, recent events need to be assessed within the context of developments since the early 1990s. The chapter shows that all Russian uses of military force throughout the post-Soviet years have been driven by a combination of motivations and none can be explained by a simple formula. These motivations are closely linked to the various roles of the military in Russian foreign policy as outlined in chapter 1, including status concerns, imperial legacy, sovereignty and insecurity, as well as multilateralism. Russia has become more assertive in the pursuit of an independent foreign policy in recent years, both vis-à-vis the former Soviet region and the West. Its military has also become more capable of achieving various foreign policy objectives. However, there is little evidence to suggest a fundamental turnaround in Moscow's views on the utility of military force. Moscow's desire to recreate the Soviet Union has never been a major motivation for interventions in the CIS region and Russia's view of the West is not as a black and white as often asserted. Russia's relationship with the West, in the past as today, is characterized by a complex interplay of conflict and cooperation.

Russian 'peacekeeping' operations in the CIS region during the 1990s

This section identifies Russia's motivations for using military force in its 'near abroad' throughout the 1990s. Tracing possible patterns in Moscow's reasons for intervening there in the past enables a more contextualized understanding of the annexation of Crimea in 2014 and of its possible implications. The section shows that in all cases of Russian military operations in the region during the first decade of the post-Soviet era, the decision to use force was determined by a confluence of factors. Although the country's imperial legacy and status concerns have certainly played a role, chance and contingency, strategic interests and insecurity have also been important. Russia's views on its interests and position in the CIS region became more clearly articulated throughout the 1990s. However, there is no evidence of a sudden 'revanchism' or the intention to use military force purely for the expansion of territory.

Imperial legacy

Russian military involvement in the 'near abroad' – first in the ethnic conflict in Moldova (Transnistria), then South Ossetia, Abkhazia (Georgia) and Tajikistan – started almost immediately after the collapse of the Soviet Union. Barely a year into the post-Soviet period, concerns that these operations were potentially driven by an imperialist agenda started being voiced (Crow 1992; Lepingwell 1994: 70). Reflecting the same interpretation of Russian actions that echoed strongly in the aftermath of the annexation of Crimea in 2014, Zbigniew Brzezinski wrote the following about Russian foreign policy already twenty years earlier: 'regrettably, the imperial impulse remains strong and even appears to be strengthening. This is not only a matter of political rhetoric. Particularly troubling is the growing assertiveness of the Russian military in the effort to regain control over the old Soviet empire' (1994: 72).

As shown in more detail in chapter 1, the implications of Russia's imperial legacy and its role in the country's foreign and defence policy are far from straightforward. When the Soviet Union collapsed, Russia's future part in the affairs of the region was uncertain and subject to significant domestic disagreement. Although some in the political elite believed from the outset that it was Russia's historical destiny to be the dominant actor, others argued that pursuing any form of reintegration was not in the country's national interest, economically or politically (Babak 2000: 93–4). Views in Moscow on the use of military force across the territory of the Commonwealth of Independent States (CIS) also differed widely during those early years. Communist and nationalist politicians believed that unilateral military force should be employed to ensure Russia's dominant position. Liberal forces, in contrast, held that, although the country had a responsibility to contain violent conflicts in its neighbourhood, it should do so only in cooperation with regional and international organizations and in strict adherence to international law. The middle ground was occupied by what Allison termed the 'moderate nationalists'. These believed that Russia had a special role as the guarantor of security in the region. If required, the country should be able to carry out its responsibilities unilaterally, but preferably through multilateral organizations for reasons of legitimacy and financial burden sharing. From 1993 onwards the moderate nationalist view emerged as a broad consensus (Allison 2013a: 122–3).

The foreign policy concept adopted in 1993 laid out Russia's views on what it saw as its interests, rights and responsibilities as the dominant security provider in the region. According to the concept, 'the objective reality is this: reciprocal interests are involved here, and – while agreeing to compromises to resolve emerging problems – Russia will not make unilateral concessions in order to develop relations; it will not accept damage to its national interests, or encroachment on the rights of ethnic Russians abroad' (Foreign Policy Conception of the Russian Federation 1993: 35). There was a clear expectation that Russia's privileged position should also be acknowledged internationally. As Yeltsin noted in 1993, 'Russia continues to have a vital interest in the cessation of all armed conflict in the territory of the former USSR. Moreover, the world community is increasingly coming to realize our country's special responsibility in this difficult matter. I believe the time has come . . . to grant Russia special powers as a guarantor of peace and stability in this region' (cited in Page 1994: 800). The view that, owing to historical legacies, CIS territory remained Russia's zone of special interest and responsibility has been an unwavering factor in the country's foreign policy since the early 1990s. As such, this view has influenced decisions to use military force there and it has also been used, in some cases, to officially justify Russian military interventions. For example, in 1993 Foreign Minister Andrei Kozyrev explained Russia's involvement in Tajikistan with reference to 'Russia's age-old historical ties, its great-power status and its vital security interests' (Page 1994: 804). The belief in Russia's privileged position in the 'near abroad' has been a constant feature in Russian foreign policy throughout the post-Soviet years and was reconfirmed in the 2014 military doctrine (Sinovets and Renz 2015).

A factor related to Russia's imperial legacy that informed discussions on the use of force in the early 1990s were the fate and rights of the millions of ethnic Russians living, as a result of the migration policies of the Russian empire and the Soviet Union, in the other newly independent states. There was a consensus among the Russian political elite at the time that there was a real threat to Russian minorities in the former Soviet region (Light 1996: 60). The need to protect the rights of ethnic Russians and Russian speakers beyond the country's own borders was present in foreign policy discussions as early as 1992 and enshrined in the foreign policy concept and military doctrine from 1993 onwards. Yeltsin and Kozyrev did not rule out the use of military force for the protection of Russians abroad (Sergunin 2007: 55). The fact that the majority population of Transnistria is Russian-speaking very likely increased sympathies to the breakaway region's cause. During a visit in the early 1990s, then-Vice President Aleksandr Rutskoi, who took a particularly hard line on the need to protect Russian speakers there, called it 'a small part of Russia' (King 1994: 112). Even with regard to the war in Tajikistan, Kozyrev noted that 'we must also remember that Russians live there' when he explained Russia's responsibility to 'act as peacemaker' in the country (Rotar 1992). However, the issue was never used formally to justify military interventions in the early 1990s, instead playing a prominent role in domestic discussions and diplomatic efforts (Allison 2014: 1283; J. Smith 2016: 47). The level of rhetorical emphasis on the fate of ethnic Russians/Russian speakers abroad increased steadily since the early 2000s and especially following the turn towards a more strident Russian nationalism in subsequent years. As we will see in the discussion of the war in Georgia in 2008 and the annexation of Crimea below, however, the degree to which concerns over the fate of Russian speakers have been driving Moscow's decisions to use military force in the 'near abroad' should not be overstated.

Chaos, chance and contingency

It is clear that imperial legacy influenced Russian uses of military force across the CIS region in the 1990s. However, it was not the only reason determining its involvement in these conflicts. It is important to bear in mind that Russian military operations in Transnistria, Abkhazia, South Ossetia and Tajikistan were not the result of a clear strategic vision, but were 'driven by events on the ground', as Allison put it (2013a: 122). In the immediate period following the collapse of the Soviet Union, Russia was as unprepared as the other newly independent states, militarily, doctrinally and politically, to deal with the multitude of security challenges created by the rapid disintegration of the former superpower. Until a clearer policy line towards the region became apparent towards the end of 1993 as discussed above, this unpreparedness led to what Pavel Baev called the serious 'muddle and disorganization' of early Russian military operations (1996: 35). The Ministry of Defence of the Russian Federation was not created until May 1992, which meant that command and control over troops located within and outside of Russian territory was weak (Hopf 2005: 230). There was no clear guidance on policy towards the CIS from the Foreign Ministry, which also had only recently been established. Coordination between both ministries was extremely limited. This led to incoherence in decision making and implementation, with the Foreign Ministry often advocating a softer response to conflict resolution, whilst some officers in the Ministry of Defence advocated a more forceful approach (Baev 1996: 33).

Soon after the disintegration of the Soviet Union, Russian troops still stationed in Moldova, Georgia and Tajikistan came under fire from local paramilitary forces, who were often demanding weapons. This further complicated a coordinated response. The situation became so tense that in June 1992, Russian defence minister Grachev reportedly 'issued an order allowing units based in "hot spots" to defend themselves without consulting with Moscow each time' (Taylor 2003: 273). As a result, Russian soldiers became party to a number of conflicts by default and the line between official policy and the initiative of local military commanders difficult to distinguish. For example, the exact circumstances and motivations for Russia's initial involvement in Transnistria are unclear until this day. According to Lance Davies, 'there is no definite evidence to suggest that Russian military units received direct orders from Moscow' (2015: 90). It is therefore possible that, in the absence of clear central guidance and policies at the time, the 14th Army under the leadership of General Alexander Lebed started intervening on the side of the Transnistrian separatists independently, rather than under the direction of Moscow. As Baev noted, by 1993 the situation had come to the point where the Kremlin had no choice but to 'adjust its course according to the military realities – or rather, to the realities created by the military' (1996: 35).

Chaos in decision making and contingency drew Russia into a variety of military conflicts during the early 1990s. No matter how chaotic and unplanned these early military interventions were, however, they set the path for the country's future role in the CIS region. In all conflicts concerned, Russian military action was instrumental in the cessation of the 'hot' phase of the civil wars. At the same time, in all cases Russia also took the chance of maintaining a lasting military presence in these countries. This not only gave it an important foothold in strategically significant 'outposts', such as the Caucasus; it also offered a powerful lever of political influence. As Vladimir Babak cautioned in 2000, 'Russia will try to use its military presence in some CIS countries (Georgia, Moldova, and some others) to pressure these states and to interfere in their internal affairs. Despite Moscow's declaration that it is ready to withdraw its military forces from CIS states, it is likely that Russia will not hurry to do so' (2000: 102).

The continuing presence of Russian troops in all cases and the 'heating up' of the frozen conflict in South Ossetia in 2008, indicates that such concerns were justified. Having said this, in addition to offering the chance of cementing its dominant role in the region, Russia's early military operations also had negative consequences for its position there today. Russian partiality in peacekeeping operations and its continued military presence in various CIS countries not only caused lasting problems in its relations with some of the countries affected directly; it also damaged Russia's image as a potential partner for multilateral security cooperation, especially amongst those CIS states that had feared Russian dominance since 1991. First impressions contributed to the fact that Russia's desire to be at the centre of a sphere of influence based not only on fear, but also on respect, has become next to impossible today. As Dmitri Trenin wrote in 2016, 'Russia's "sphere of influence" is actually limited to the territories it physically sustains and protects: Abkhazia; Donetsk and Lugansk; South Ossetia; and Transnistria' (2016b: 42). Inasmuch as having a sphere of influence is an important characteristic of a true great power, Russian military interventions in the CIS region since the early 1990s have had significant negative consequences for the achievement of one of its central foreign policy goals.

Strategic vulnerability and insecurity

In addition to imperial legacy, chance and contingency, there were other factors making Russia's prominence in the region almost a foregone conclusion. A sense of strategic vulnerability and insecurity certainly played a role in this respect. Russia's strategic and security interests in the region did not simply dissipate with the collapse of the Soviet Union. Interdependence with the other newly independent states as a result of the shared Soviet past, including in the military realm, meant that the option of abandoning all interests and responsibilities in the region was simply unrealistic. For example, when the violent clashes in Transnistria developed into open military conflict in early 1992, around 23,000 Russian troops were still stationed on Moldovan territory. This in itself meant that Russia had a strategic interest in the situation there. When the idea of preserving joint military forces covering the entirety of the CIS was abandoned and Russia created its own national military forces, many hundreds of thousands of its service personnel were stationed across the territories of the other newly independent states. The process of withdrawing these forces, with the exception of the limited foreign deployments that remained in a number of CIS states, lasted until well into the 1990s (Trenin 2011: 76). Moreover, as discussed in chapter 2, important material assets were suddenly located in foreign countries, because the bulk of significant military installations had been located at the periphery of the Soviet Union. Disputes over the ownership of Soviet military hardware and installations were sensitive and could not be resolved overnight. This necessitated Russia's ongoing investment in the security affairs of the other CIS states for years to come. In the event, Russia forfeited many of its prized assets, especially in Ukraine and in Belarus (Lambeth 1995: 88). However, the dispute with Ukraine over the Black Sea Fleet and Sevastopol naval base, which were particularly important strategic objects for Russia, was not resolved until 1997 (Taylor 2003: 274).

An issue that made a dominant Russian role in conflict management across the CIS almost inevitable was the fact that, in spite of all the problems its own armed forces faced, it was the only country during the early post-Soviet years that had anything approximating a functioning military. With the exception of Ukraine, none of the other newly independent states had the capacity to create capable national forces from the assets they had inherited from the Soviet military and they remained dependent on Russia for years to come (Allison 1993a: 4). This meant that, although attempts were made to establish multilateral cooperation in the management of the ethnic conflicts erupting in the early post-Soviet years, lack of capacity of the other newly independent states meant that CIS peacekeeping efforts were inevitably dependent on Russian troops and materiel. Moreover, not all of the CIS states saw Russia as a threat and some even looked towards Moscow as a provider of security. Particularly Tajikistan, ravaged by civil war from autumn 1992, welcomed Russian military involvement as this was seen not as a danger to, but as a guarantee of, its new statehood and the only possibility for the survival of President Emomali Rakhmonov's regime (Olcott 1995: 361). Even Georgia, which remained outside of the CIS until the end of 1993 and had a difficult relationship with Russia due to the latter's siding with Ossetian and Abkhazian separatists, initially pondered the formation of a strategic axis with Russia, because it recognized that it was not in a position to protect the country's new statehood independently (Allison 1993a: 64–5).

Perceived threats to sovereignty were an important source of insecurity and reason for Russia's involvement in ethnic conflicts in the CIS since the early 1990s. The conflicts in Abkhazia and South Ossetia stoked fears in Moscow about escalation and the potential overspill of hostilities and instability into the Russian side of the North Caucasus, which already was a troubled region. As Sergei Markedonov (2007) noted, 'security in Russia's Caucasus is impossible without stability in Georgia'. Concerns over instability in this region as a threat to Russian territorial integrity led to the internal use of military force, the Chechen War, in 1994. Tajikistan is geographically remote from Russia, but the civil war raging on in the country from 1992 quickly turned into a Russian and CIS-wide source of insecurity (Allison 1993a: 54–5). A particular worry was Tajikistan's long and poorly defended border with Afghanistan. This exacerbated the potential for lawlessness, crime, insurgency and extremism as a significant threat to the security of other Central Asian states and beyond (Gleason 2001: 80–3). Although the operations in Tajikistan were couched in terms of a CIS peacekeeping operation, they were never neutral inasmuch as both Russian and Uzbek contingents supported the Rakhmonov regime. The consolidation of power under his government was seen as the most hopeful path towards stability in the country, which was a priority for both Russia and other Central Asian states. The civil war in Tajikistan came to an end in 1997. However, ongoing concern over the spillover of lawlessness as a source of instability in the region meant that Russian border guards continued to secure the country's borders with Afghanistan until 2005 (Blua 2004; McDermott 2005).

Russian military power and the West during the 1990s

The Russian Federation has never engaged in offensive military action against a Western state. This does not mean, however, that military power has not been important in its foreign policy beyond the borders of the former Soviet Union. Although little attention was paid to the significance of military power in Russia's interaction with the West until the annexation of Crimea in 2014, it has played a role in this respect since the early 1990s. As explained in the book's introduction and in chapter 1, the utility of military power is not limited to the fighting of wars and defeating of opponents. Instead, it is routinely wielded by states, including by Russia, in a variety of other physical and symbolic ways for the achievement of specific foreign policy objectives. Tracing the role of military power as a tool in Russian policy towards the West since the beginning of the 1990s is important for a contextualized reading of what the operations in Crimea and Syria signify for East–West relations.

Russia's actions in both cases have been interpreted as a threat to the West inasmuch as they implied the Kremlin's preparedness for uncompromising military confrontation in pursuit of great power status. The following section shows that the use of military power in Russia's quest for great power recognition dates back to the 1990s. It also argues, however, that its function in this capacity has not been straightforward. The peacekeeping operations in the Balkans in the mid-1990s and developments relating to the Kosovo War in 1999 show that Moscow does not view the value of military power exclusively as the ability to fight wars. Instead, Russia perceives security cooperation with other major powers as essential for the maintenance of its own great power status. Both examples demonstrate the complex interplay of conflict and cooperation in Russia's relations with the West. An appreciation of this is essential for an informed understanding of more recent events.

Russia's involvement in UN peacekeeping and the conflicts in the Balkans

Status concerns have influenced Russia's relations with the West since the collapse of the Soviet Union. In the early post-Cold War years, Russian foreign policy was strongly Western-centric. In the West, this outlook was interpreted as Moscow's readiness to become a part of the West, following a period of transition assisted by its new partners. In contrast, Russia never saw closer cooperation with the West as a means for full integration, but for the maintenance of its great power status. As Andrew Monaghan has noted, this dissonance in understandings has shaped the East–West relationship until today (2016: 60). While economic considerations, such as the need to integrate into the global economy and the hope for financial assistance, figured in Russia's desire to establish closer relations with the West, political consid-erations were also significant. Building on Gorbachev's New Thinking, which sought to define a new role for Russia as an influential global actor in spite of its weaknesses at the time, the political leadership in the early 1990s believed that cooperation with multilateral institu-tions and the pursuit of a 'strategic partnership' with the West was the best way to ensure the country's security and enduring relevance as an international actor (Tsygankov 1997: 249–50; Facon 2006: 37). As a great power, Russia believed that its responsibilities extended beyond its immediate neighbourhood. As such, it had the duty to contribute to security and stability on an international level in cooperation with other major powers (Headley 2003: 210).

Russia's involvement in UN peacekeeping activities was seen as an important means for integration into multilateral security organizations. Although the Soviet Union had gained limited experience in UN peacekeeping since 1973, the country's contributions to UN missions had significantly increased under Gorbachev. This trend continued into the post-Soviet years. Russian peacekeepers participated actively in the missions in the Balkans from 1992 onwards, contributing up to 1,500 troops at a time to various protection and stabilization forces (UNPROFOR, IFOR, SFOR) and working alongside NATO troops as part of KFOR in Kosovo until 2003. Building a relationship with the West and finding its place in the post-Cold War order was an important motivation for Russia's involvement. As Alexander Nikitin (2004) wrote about Russia's decision to deploy a brigade to IFOR, this 'was not taken simply to help rebuild stability in Bosnia and Herzegovina, but has to be seen within the context of relations between Russia and the West in the post-Cold War era'.

Peacekeeping in the Balkans differed significantly from the experience in the CIS, where Russia acted mostly on its own and was criticized for lack of adherence to traditional peacekeeping principles. As the missions in the Balkans were led by NATO, albeit under a UN umbrella, Russia's participation was not free from friction. However, precisely because of the role played by NATO, the involvement of Russian peacekeepers was seen as an important precedent both in Russia and in the West, as it offered the potential for closer cooperation in the future. Experience gained in the Balkans brought Russian peace-keeping doctrines more into line with the views of the UN and of the West and also informed reforms of the security apparatus (Facon 2006: 39). Practical cooperation improved the interoperability of NATO and Russian forces and led to the development of a Generic Concept of Joint NATO–Russian Peacekeeping Operations (Nikitin 2004).

In spite of the relative success of multilateral peacekeeping in the Balkans, the rift in the respective understandings by Russia and the West of the exact nature of their new partnership soon started to make itself felt. As discussed in chapter 1, even when Russia was severely weakened in the early 1990s, the idea that the country could be anything but a great power was inconceivable. As such, Moscow expected to be given an equal voice in multilateral decision making on questions of international security – an expectation that, in the eyes of the Russian leadership, was bitterly disappointed. In an article published in *Foreign Affairs* in 1994, the Russian foreign minister, Andrei Kosyrev, outlined what he saw as the differences in Russia's and the West's approaches to cooperation. In February 1994 NATO had issued an ultimatum and threatened air strikes on Bosnian Serb positions following an attack on a marketplace in Sarajevo without consulting Russia. This led to outrage in the country's diplomatic and political circles. Referring to this episode, Kosyrev stated that Moscow was not prepared to accept a subordinate place in global affairs: 'the international order in the 21st century will not be Pax Americana . . . The United States does not have the capability to rule alone. Russia, while in a period of transitional difficulties, retains the inherent characteristics of a great power (technology, resources, weaponry)' (1994: 63). Repeatedly asserting Russia's commitment to a relationship based on mutual trust and respect, Kosyrev criticized the West for acting on the basis of 'paternalism and assumed inequality'. In order to avoid renewed conflict, he noted that 'we should decide what we mean by partnership: close and sincere cooperation in world affairs or a lopsided relationship in which all rights are on one side and all obligations on the other?' (1994: 62, 64). Status concerns and the failure of multilateral cooperation, as Russia saw, it had led to the first serious crisis in East–West relations in the post-Soviet years (Headley 2003).

The Kosovo War in 1999

The significance of NATO's Operation Allied Force (OAF) as a key event in the emergence of a more assertive Russian foreign policy and a factor shaping Moscow's thinking on the utility of military force cannot be overstated. Understanding the reasons for why OAF had such a strong impact on Russian foreign and defence policy thinking is important for a contextualized analysis of more recent events.

Russia's inability to prevent NATO's airstrikes against Serbia, which it vehemently opposed, reinforced concerns over the country's international status and painfully highlighted its limited clout in shaping global affairs (Averre 2009: 580; Allison 2013a: 44). The Kremlin's failure to influence events diplomatically underlined that the country's great power status could not be maintained without strong military forces. During OAF, the Russian executive came under increasing pressure to respond militarily to NATO's actions. In April 1999, the State Duma voted strongly in favour of supporting Milosevic with weaponry and military advisers. The head of the Ministry of Defence's Main Directorate for International Cooperation, General Leonid Ivashov, publicly stated that the Russian armed forces were ready for joint operations with Yugoslav forces if called on to do so (Cross 2002: 2–3). Yeltsin did not heed these calls, but he warned that NATO should take care not to push Russia into using military force (Averre 2009: 579–80). Given the weakness of Russia's armed forces, unilateral military action in support of Milosevic was not within the realm of possibilities at the time. In June 1999, a few hundred Russian peacekeepers were deployed from Bosnia to seize Pristina airport. The impact of this on the conduct of OAF was negligible. However, the action sent a signal, within the constraints of Russian military power at the time, that the Kremlin was not prepared to stand by and accept its complete exclusion from the resolution of the conflict. The dash to Pristina sped up negotiations about the involvement of Russian peacekeepers in KFOR (Antonenko 1999: 138). However, Russia's inability to stand up to NATO was seen as a humiliation and betrayal of the country's military by many in the political elite and defence establishment (Brovkin 1999: 557). When Putin rose to political prominence in the autumn of 1999, efforts to return the Russian military to its former glory took an important place in his agenda from the beginning. OAF marked a watershed in Russian views on its defence requirements, as discussed in chapter 2. It was the beginning of the military revival as an important element in the restoration of what Putin called 'the country's prestige and leading role in the world' (Putin 2000).

Military weakness, preventing Moscow from shaping international events at the time of the Kosovo War, also heightened concerns over the country's sovereignty. NATO's ability to pursue OAF without a UN Security Council resolution and in spite of strong Russian opposition, suggested that the West had a degree of sovereignty in foreign policy making that Russia lacked. The fact that OAF was directed against a Russian ally was significant in this respect. Although ethnic kinship and historical ties with Serbia did not necessarily determine Moscow's strong reaction to the war, its affinity with Milosevic's Yugoslavia, which looked up to Russia as a powerful friend and ally, was certainly a factor. As Sharyl Cross has put it, 'the fact that Russian assistance was actively solicited by Serbs reaffirmed for the Russians recognition of their continued importance' (2002: 15). Lacking the military power required to render assistance to its ally in this case put into question the country's foreign policy sovereignty. It also stoked fears that, unless its military weakness was dealt with, Moscow would not be able to prevent potential intrusions by the West into its more direct sphere of influence, the CIS, in the future (Allison 2013a: 44).

NATO's justification of OAF as a humanitarian intervention was a particular bone of contention and exacerbated fears over the country's sovereignty. Although Russian attitudes towards humanitarian intervention are not straightforward, and its principles as such are not necessarily rejected, the Kosovo War marked the beginning of Moscow's strong opposition to what it sees as the West's means of implementing the concept (Averre and Davies 2015: 814). Moscow had argued that decisions on action to be taken against Milosevic should be made by the UN and authorized by the Security Council, which included Russia and other important powers. When NATO proceeded without a Security Council Resolution, Moscow viewed this as the West's blatant disregard of international law and of Westphalian norms. As Vladimir Baranovsky noted, it was OAF, much more so than the process of enlargement, that consolidated Russia's negative views of NATO. The forceful intervention against Milosevic's regime confirmed to many in Russia the alliance's 'aggressive character' (2000: 115–16). The Kremlin's belief that the West abuses humanitarian principles as a pretext to intervene in the domestic affairs of sovereign states in order to expand its own influence emerged. In addition to concerns over the potential for NATO's use of such justifications to push into the CIS region, there were popular fears that the 'Kosovo model' could ultimately be applied to Russia, especially as evidence of human rights abuses in Chechnya mounted. Although it is unlikely that the Kremlin believed such a scenario was imminent, Serbia's inability to deter outside intervention underlined the need for a strong military (Allison 2013a: 57).

The Kosovo War also left Russia feeling less constrained in the use of military force in pursuit of its national interests. In the Kremlin's eyes, what it saw as NATO's flagrant violation of international law and flimsy justification for aggressive military action, invalidated any Western legal and moral criticisms of its own conduct. In this sense, as Russia saw it, Kosovo made the use of force 'less unjustifiable'. As Baranovsky explained: 'NATO has set a precedent, and Russia should not hesitate in the event that it considers the resort to military means necessary' – 'we had to "swallow" your actions in Kosovo, you have to do the same with respect to ours' (2000: 124–5). As we will see in the discussion of the war in Georgia in 2008 and Ukraine in 2014, Russia justified its own operations there on humanitarian grounds. This has been flagged up as a contradiction in Russian foreign policy, for example by Derek Averre, who described Russian explanations for the war in Georgia as 'an almost bizarre reversal of its reaction to OAF' (2009: 590). The fact of the matter is that Russia did not expect the West to accept these explanations. The need for Western approval of its actions simply did not factor into calculations. The Kosovo War made Russia more assertive in pursuing objectives that do not coincide with Western interests. It confirmed to the Kremlin that 'Russia has to rely on military strength, rather than on illusions about justice and good intentions in international relations' (Baranovsky 2000: 124–5).

The crisis in the Balkans offered an opportunity to establish closer cooperation between Russia and the West in solving global security issues. As a result of dissonant views over the nature of Russia's position in the post-Cold War international system, the crisis ultimately led to the first serious breakdown in relations. Even at the height of tensions over OAF, however, the Kremlin sought cooperation with the West. Following a brief suspension of relations with NATO, cooperation was resumed and Russian peacekeepers were involved in KFOR until 2003. The tensions over Kosovo, as far as Moscow was concerned, were caused by the West's refusal to give Russia a say in the resolution of the conflict which, as a great power, it thought it deserved. The feeling of being sidelined in decision making led to a more confrontational approach towards the West, including the deployment of Russian soldiers to Pristina airport. The decision to use military power in this case was determined by Moscow's desire for inclusion. The Kosovo War demonstrates the difficult interplay of conflict and cooperation in Russia's relations with the West. The ability to pursue an independent foreign policy and to shape events of international significance, even if this leads to conflict with the West in certain cases, is required if Russia is to gain great power recognition. As Tsygankov noted, however, the pursuit of 'great power status is not a goal in itself . . . but a necessary condition for more advanced engagement with the world' (2008: 49).

The war in Georgia, 2008

Understanding the reasons for Moscow's decision to use military force against Georgia during a five-day war in August 2008 is important for a contextualized analysis of more recent operations. On the one hand, the war's roots are in the early post-Soviet years, when Russia first became involved in the ethnic conflicts in South Ossetia and Abkhazia as discussed above. Echoing views of Moscow's foreign policy towards the 'near abroad' dating back to the 1990s, many Western observers interpreted the 2008 events as evidence of renewed Russian imperialism (Rich 2009: 242). On the other hand, in comparison to the CIS 'peacekeeping' operations during the first decade of the post-Cold War years, which were widely seen as a regional issue, the 2008 war in Georgia also had a distinctly international dimension. It occurred against the background of an increasingly confrontational tone in Russian foreign policy rhetoric towards the West (Stent 2009: 1090). As such, another popular interpretation of the conflict was that Russia was fighting a proxy war against the West and was intent on instigating a New Cold War (Rich 2009: 241). As Peter Shearman and Matthew Sussex argued, the interpretation of Russian actions in Georgia in 2008 as 'aggressively unilateral' and as the product of imperialism and/or new cold wars is flawed and does not reflect the 'convergence of several distinct but interlinked threats to Russian interests that Moscow had consistently articulated in the post-bipolar security environment' (2009: 252, 270). An appreciation of the factors determining Russia's decision to use force against Georgia in 2008 is required for a more informed assessment of continuity and change in Moscow's view of military power as a tool of foreign policy.

It is important to note that chance and contingency played a role in the 2008 war against Georgia. In other words, Russian military actions did not come out of the blue, but were the result of a specific confluence of circumstances. By 2008, Russian peacekeeping forces were still stationed in both South Ossetia and Abkhazia. Ongoing instability in the North Caucasus meant that the region was as strategically important to Moscow as ever. Russia's relations with Georgia had never been easy, but following Mikheil Saakashvili's election as president in 2004, attitudes had turned increasingly bitter on both sides. Georgia became more vocal about its unhappiness with Russia's continued military presence and Saakashvili vowed to restore the country's control over the disputed territories. Russia in turn criticized what is saw as 'Georgian bellicosity toward South Ossetia' (Tsygankov and Tarver-Wahlquist 2009: 307). A steady build-up of Russian military activities in the North Caucasus and near the Georgian border from early 2008 indicate that Russia both expected and planned for the escalation of these tensions (Kramer 2008: 7; Vendil Pallin and Westerlund 2009: 400). However, it was the shelling of the South Ossetian capital Tskhinvali by Georgian artillery and the death of civilians and Russian peacekeepers, which gave Moscow a reason to intervene.

Self-defence was at the core of Russian justifications for the use of military force. Given that hostilities were initiated by Georgia, Russia had some legitimate ground on which to base this argument. This was also acknowledged in the 'Tagliavini report', the findings by a fact-finding mission established by the European Union (Independent International Fact-Finding Mission 2009). It is notable that Russia also referred to humanitarian norms, arguing that Georgian atrocities against civilians necessitated military intervention. As noted above, this should not be read as a turnaround in Russian views on humanitarian intervention. As Allison showed, there was no evidence of systematic genocide committed by the Georgian side, and the disproportionality of Russian responses also violated humanitarian principles. Humanitarian concerns were not a reason for Russian actions in Georgia, but a rhetorical device used for instrumental purposes (2008: 1152–3). Just as NATO had subverted humanitarian norms in pursuit of its own interests in Kosovo, as Russia saw it, Moscow manipulated them to suit its own purposes in Georgia (Kernen and Sussex 2012: 92, 111).

Insecurity and self-defence do not explain the scale of Russia's response and why military actions extended significantly into undisputed Georgian territory. This is because regional status concerns were also important in Russia's decision to use military force. As Georgia's relations with Russia deteriorated, Saakashvili pursued an openly pro-Western foreign policy, including close bilateral relations with the United States and the long-term goal of joining NATO. In 2006, Georgia had withdrawn from the CIS Council of Defence Ministers, announcing that, as a future member of NATO, it could not be part of two rival military alliances simultaneously (Kramer 2008: 7). In its efforts to forge closer relations with the US, Georgia also contributed troops to the war in Iraq. These developments suggested to Russia that losing the country as a part of its 'sphere of influence' was a distinct possibility. The prospect of 'losing' Georgia is unacceptable to Russia, because of its strategic and security interests in the region, including in the realm of energy. Cultural and historical affinity are also important. In this sense, Russia's imperial legacy factored into the decision to use force in this case (Shearman and Sussex 2009).

International status concerns were also important. Since the Rose Revolution in 2003, which like subsequent 'colour revolutions' was interpreted by the Kremlin as a tool used by the West to extend its power, Russia saw Georgia as an important locale for potential intrusion into its perceived sphere of influence. Moscow therefore viewed the courtship between Georgia and the West as mutual. In 2006, Abkhaz and Russian officials claimed that Georgia was acting in the interest of the West and pursued NATO membership so the US could open military bases in the Caucasus. Russian fears over the West's 'intrusion' into Georgia were aggravated when, during a visit to Tbilisi in spring 2008, US Secretary of State Condoleezza Rice voiced her support for a Georgian NATO Membership Action Plan, at the same time as criticizing Russian conduct in South Ossetia (Shearman and Sussex 2009: 257).

The prospect of Georgia's NATO membership and its relationship with the West never figured in official Russian justifications for the August 2008 intervention. However, events predating the operation suggest that this was a factor. During the NATO summit in Bucharest in April 2008, the alliance officially 'welcome[d] Ukraine's and Georgia's Euro-Atlantic aspirations for membership in NATO' and stated officially that both states will become members of the alliance in the future (NATO 2008). Following a meeting of the NATO–Russia Council at the same summit, Putin reiterated that NATO enlargement would be viewed as a 'direct threat' to the security of Russia (Putin 2008). The August 2008 war made Georgia's membership in NATO in the immediate future unlikely. As such, weakening Georgia was 'not just a goal but an *instrument* for Russia', as Allison put it, in the pursuit of higher-order foreign policy objectives (2008: 1165). Commenting on Russia's global vision in the aftermath of the war in Georgia, President Medvedev reiterated that Russia's dominant status in the 'near abroad' was not negotiable: 'domination is something we cannot allow . . . there are regions in which Russia has privileged interests . . . We will pay particular attention to our work in these regions . . . As for the future, it depends not only on us but also on our friends and partners in the international community. They have a choice' (Medvedev 2008).

Finally, the war in Georgia confirmed the ongoing dissonance in Russian and Western views on the nature of their relationship, as well as the complex interplay of conflict and cooperation in Moscow's approach to the West. Although the war in Georgia was consistent with the interests and ambitions Russia had articulated for many years, it was met with surprise in the West. This was because, although Putin's foreign policy rhetoric had been at times confrontational, there had been the perception of a general improvement in relations (Rich 2009: 240). Russia had been open to security cooperation with the West since 2000, including with NATO. This included participation in the efforts to deal with 'new security challenges' in a multilateral setting, as discussed in the previous chapter, but also more 'traditional' concerns. Following the 9/11 attacks, Russia pledged moral and material support to the US in the Global War on Terrorism. It did not withdraw its cooperation even in moments of extreme tension, such as the start of the Iraq War in 2003, which the Kremlin vehemently opposed. This 'realignment' with the West in the aftermath of 9/11 was read by some commentators as a sign that Russia was finally ready to progress 'in the 21st century as an integrated member of the Euro-Atlantic community' (Cross 2006: 175). Although Russian great power aspirations remained an obstacle, as Angela Stent and Lilia Shevtsova wrote in 2002, this could be overcome by a combination of *realpolitik* and democracy promotion on the part of the US, and Russia's 'integration with the West with full acceptance of liberal democracy' (2002: 132).

Such interpretations of Russian actions display a fundamental misunderstanding of Moscow's approach to the West, where conflict and cooperation are not mutually exclusive. Russian support for the global war on terrorism was not a sign of the Kremlin's readiness to integrate. Support was rendered, because security interests coincided, given Russia's own struggle with terrorism for many years. The war on terrorism presented an opportunity where Russia could engage in multilateral efforts to solve an important international problem together with other great powers. It also allowed it to present its own actions in Chechnya in internationally acceptable terms. Cooperation with the West in the war on terrorism did not preclude a more confrontational approach when interests diverged, as they did in Georgia in 2008. On the one hand, Russia's reassertion of its status by military means in this case inevitably caused tensions and made cooperation with the West more difficult. On the other hand, tensions with the West were not only a price worth paying, but instrumental in Russia's quest to gain respect as a great power that is able to pursue an independent foreign policy. Such respect is essential, in Russia's eyes, for cooperation with the West on an equal footing. Following the war in Georgia, Russia continued to support the NATO-led International Security Assistance Force (ISAF) in Afghanistan by providing air and land access for the transport of cargo through its territory. In 2008 it also agreed to become an official part of NATO's Northern Distribution Network, which included a trans-shipment hub at Ulyanovsk in the Russian heartland (Boguslavskaya 2015: 213–15).

The annexation of Crimea and war in Ukraine, 2014

Russian military actions in Ukraine, and the annexation of Crimea in particular, have been interpreted as a 'paradigm shift' in Moscow's foreign policy and as evidence of a 'seismic change in Russia's role in the world' (Rutland 2014). The assessment of these actions as a dramatic and sudden turnaround was based on the fact that, for the first time since the creation of the Russian Federation, the country grabbed a piece of another sovereign state's territory. This seemed to suggest a qualitative change in the Kremlin's perceptions of its historical rights and responsibilities in the 'near abroad' from more indirect forms of domination to an expansionist vision. It caused concerns that further expansion of territory into the CIS region and even beyond was likely and had to be deterred. The use of force for territorial expansion was certainly new inasmuch as it had not previously featured in Russian military operations during the post-Soviet years. The clear illegality of this move also makes the focus on this aspect understandable. However, the assumption that the acquisition of territory was Russia's foremost motivation for the use of military force in this case represents a limited explanation of the war in Ukraine. A more contextualized understanding of Russian motivations is required not to justify these actions, which are in any case unjustifiable. Rather, it is needed to enable the identification of responses that will not inadvertently cause tensions to spiral.

The annexation of Crimea did not occur in a vacuum. Developments leading up to the event indicate that Russian actions were not the result of a 'paradigm shift', but a continuation of interests and threat perceptions that had driven Moscow's foreign policy for a long time. As was the case in Georgia in 2008, regional status concerns were significant. Fears in Russia over its waning influence over Ukraine date back to the Orange Revolution in 2004. This had brought to power a political regime in Ukraine that questioned what Russia saw as the status quo in the region. Although the new leaders' Western leanings were expressed more strongly in rhetoric than in actual policies, they advocated NATO and EU membership and there was a sharp swing in foreign policy towards the West. The Orange Revolution was followed by a period of Russian coercive economic diplomacy in its efforts to bring Ukraine back into line (Charap and Colton 2017: 74–81). When Ukraine elected Viktor Yanukovich, a Russian-friendly politician, as president in 2010, Moscow's most acute geopolitical concern – the prospect of Ukraine's membership in NATO, however remote – was alleviated. However, the new leadership did not decisively reorient its foreign policy towards Russia and also continued cementing closer relations with the West, and especially with the European Union as a major trading partner. Preparations for an association agreement (AA) on trade with the EU proceeded. Russia in turn sought to enforce Ukraine's reintegration into its own 'sphere of influence'. It importuned the leadership with sticks and carrots to join the Customs Union, which it had formed with Belarus and Kazakhstan in 2009. Much political see-sawing followed as the Ukrainian leadership decided where the country's priorities should lie. In autumn 2013, Yanukovich abandoned imminent plans to sign the AA at short notice, a decision that was met with significant economic rewards by Moscow. Demonstrations in Kyiv followed and quickly turned into demands for Yanukovich to go. In spite of, or perhaps because of, the brutal suppression of demonstrators, the protests gathered momentum and lasted for months. Negotiations in February 2014, which included Yanukovich and Ukrainian opposition leaders, as well as official representatives from EU countries and Russia, failed to solve the crisis. Yanukovich fled the country and was replaced by a Western-oriented government that put the EU AA back on the table (Charap and Colton 2017: 114–26). A few days later, Russian military operations in Crimea commenced.

The developments in Kyiv aggravated Moscow's fears that it was in danger of losing Ukraine as part of its 'sphere of influence'. Even the prospect of this is unacceptable for Russia. As a large country located in the westernmost part of Russia's 'near abroad', Ukraine is of particular strategic importance, forming a buffer against NATO territory in Europe's north and east. Ukraine's status as a transit state for Russian gas to lucrative Western markets is also significant. Access to Crimea is non-negotiable in the Kremlin's eyes, because the Sevastopol naval base is central for power projection in the Black Sea region and beyond. Disputes over basing rights have led to tensions with Ukraine in the past. Annexation deprived any future Ukrainian government of the opportunity to revoke them. Long periods of shared history have contributed to Russia's view of Ukraine as an indispensable part of its 'sphere of influence'. Crimea is of particular symbolic importance in this respect (Wydra 2004). As Trenin wrote in 2011, when the dissolution of the Soviet Union led to a significant reduction in the territory controlled by Moscow, 'Crimea was the only territory outside of the perimeter of the new borders of the Russian Federation about which most Russians, irrespective of their political orientation, felt strongly' (2011: 45). In this sense, imperial legacy and the view of Russia's privileged rights in the CIS region factored into the decision to use military force.

International status concerns heightened Russia's preparedness to opt for military action. As was the case in Georgia in 2008, the Kremlin acted on the assumption that political developments in Ukraine over the past decade had been steered, or at least heavily encouraged, by the West in its efforts to expand its influence into Russia's orbit in the CIS region. The EU's offer of an association agreement and discussions in NATO of the possibility of a Membership Action Plan for Ukraine following the Orange Revolution were interpreted as evidence of this. Moreover, Moscow believed that the Orange Revolution itself had been fermented by the US. This cemented its view that revolutionary change had become a tool in the West's strategic contest against Russia in the CIS region, used 'to shift local foreign and security policy alignments in their favour by replacing incumbent leaders' (Allison 2013a: 133–4). When in February 2014 the US and other Western governments officially welcomed the new Ukrainian government only a few days after the change in power had occurred, Moscow was convinced that it had yet again become the victim of a 'Western plot', this time 'to install a loyal government in Kyiv that would move Ukraine toward the EU and even NATO' (Charap and Colton 2017: 126). The importance of international status concerns in Russia's decision to use military force in Ukraine was confirmed by Putin's heavy emphasis on the West's responsibility for the events in his 'Crimea speech' in March 2014 (Putin 2014a).

Perceived threats to sovereignty also factored into the Kremlin's decision. When it comes to developments in Ukraine, long-held

Russian concerns over the West's perceived interference in the domestic affairs of states in order to expand its influence stoked anxieties in the Kremlin over the stability of its own regime. Owing to Ukraine's close geographical proximity, the dramatic political changes in the country were uncomfortably close to home. Moreover, Putin was still reeling from the memory of extensive street protests following the State Duma elections in 2011 (Allison 2014: 1289–90; Charap and Darden 2014: 10). As such, as Allison argued, the annexation of Crimea, which dramatically improved Putin's approval ratings, also served to 'harness[ing] populism for domestic regime consolidation' (2014: 1291–3).

Russia's use of military force in Ukraine did not denote a 'paradigm shift', but a continuation of drivers that had long been evident in Russian foreign policy. There is still the question whether further territorial expansion is likely. The annexation of Crimea raised fears that neighbouring countries with large Russian-speaking minorities, such as the Baltic States, were particularly threatened. This is because claims about the need to protect 'ethnic Russians, Russian citizens, Russian compatriots, Russian-speakers' were a central plank in official justifications for Moscow's actions in Ukraine (Allison 2014: 1282). Several analysts have since pointed out that, as was the case in Georgia in 2008, the protection of Russians was not a reason for the operation, but a rhetorical device, in this case to garner support amongst domestic audiences (Allison 2014: 1296; Biersack and O'Lear 2014: 252–3). In spite of expectations in spring 2014 that the annexation of further Ukrainian territories inhabited by a Russian-speaking majority was likely, this never materialized. This was not for lack of opportunity or capabilities. As discussed in chapter 1, imperialist expansion historically has been a perilous undertaking, especially if it involved the inclusion of resistant populations. As a result of the Chechen wars,

Russia is painfully aware of the costs of suppressing a drawn-out insurgency. This would be a likely requirement in most areas other than Crimea, and especially in the Baltic States. From this point of view, further territorial expansion appears at least fairly unlikely. In eastern Ukraine, Russian hostilities have resulted in a 'simmering' conflict that will in all probability turn into a 'frozen' variant over time. This offers Moscow an important lever of influence and control over Ukraine. It is likely that, in the Kremlin's eyes, the value of such a lever vastly outweighs the benefits of adding further territory and populations to an already vast state (Bukkvoll 2016a: 276; Pifer 2017).

Russian airstrikes in Syria, 2015

Russia's intervention in the Syrian civil war came as a surprise to the West, because it was the first time the country launched a sizeable and unilateral out-of-area operation. Although the airstrikes were not directed against Western forces already active in the country, the operations have been interpreted by many as an act of aggressive confrontation aimed against the West. As Angela Stent put it, Moscow's 'unanticipated military foray into Syria has transformed the civil war there into a proxy US–Russian conflict and has raised the stakes in the ongoing standoff between Moscow and Washington' (2016: 106). A contextualized assessment of Russia's decision to intervene in Syria shows that such an explanation reflects a one-sided understanding of why the Kremlin intervened.

Moscow's decision to use force in support of Syrian President Bashar Assad's regime was determined by a specific confluence of factors. The choice of Syria as the locale of Russia's first military foray beyond the former Soviet region was not accidental. Historical ties raised Russia's stakes in having a say in the outcome of the ongoing war. Relations with Syria date back to the Soviet era and Moscow maintains some material and strategic interests in the country. These include a minor naval facility in Tartus, an eavesdropping station and revenue income as Syria's major arms supplier. Although none of these interests are significant enough to merit a costly military intervention, the strategic value of Syria was enhanced, because it represented the last bastion of Russian influence in the Middle East (Allison 2013b: 800–7). Although the Kremlin's affinity with Assad himself should not be overstated as a reason for the intervention (Katz 2013: 38), the fact that Syria regarded Russia as an ally played a role. At the time of growing international isolation in the aftermath of Crimea, being petitioned for assistance by the Assad regime reaffirmed to Russia its importance as a global power.

Insecurity – globally, regionally and domestically – caused by the growing influence of radical Islamic groups in the Syrian war was another factor in Russia's decision to render military support to Assad. Moscow had long expressed concerns over the increasing prominence of such groups, including Al Qaeda and the Islamic State, in the opposition forces fighting the incumbent regime. Referring to developments in Iraq and Libya, where externally driven regime change had led to lawlessness and the spread of religious extremism, the Kremlin saw the continuation of Assad's rule as the best option to suppress such groups and to return stability to the country (Putin 2013). In Russia's eyes, the spread of extremist forces was not only a threat to the stability of Syria itself, but also to the entire Middle East. Moreover, the international reach of groups like Al Qaeda and the Islamic State meant that there was the potential for a significant increase in terrorist activities in areas far beyond the region, including in Central Asia and, ultimately, in Russia itself (Averre and Davies 2015: 820–1).

International status concerns also influenced Russia's decision to resort to military force, as the Kremlin increasingly felt that it was being sidelined by the West in multilateral efforts to resolve the crisis. External military intervention in Syria, first by a number of Western states and then by Russia, was preceded by efforts to solve the crisis through diplomacy. From 2011 until 2013, Moscow repeatedly asserted that it had no intention to enter the war in Syria and also called on the US to refrain from military action (Katz 2013). Like in the run-up to the Kosovo War, Russia insisted that the UN was the best forum for dealing with the regime in question, so major powers in the Security Council could steer the efforts to find a lasting solution that was acceptable to all. In the event, Russia blocked any initiative that involved the forceful weakening of Assad's rule or that made his departure from power a condition. As the humanitarian crisis in Syria gathered pace, and some Western leaders became more vocal in calling for Assad to go, the chances of solving the crisis through the UN became ever more remote (Allison 2013b). When the US launched airstrikes, the Kremlin saw this as a failure of multilateralism and as yet more evidence of the West's refusal to give it an equal voice in the solution of international problems. In Kosovo, Russia was not in a position to protect its ally from NATO forces and emerged from the conflict feeling humiliated. By 2015, Russia had recovered the strength required to stand up against the West. In this sense, Russia's decision to prop up the Assad regime, targeting moderate opposition groups in addition to extremist elements, had as much to do, as Charap put it, 'with anxieties about the implications of US power than it does with Syria itself' (2013: 37). Military intervention allowed Russia to demonstrate that it now had the capabilities to challenge what it saw as the West's monopoly on the use of force on a global level. It also sent a message to the rest of the world that the country's backing was yet again 'something truly worth having' (Matthews 2016). As such, the war in Syria brought Russia closer to its goal of great power recognition (Knight 2015).

Although the airstrikes in Syria showed that Russia had regained the power to pursue an independent foreign policy beyond the borders of the CIS region, concerns over domestic sovereignty also informed the decision to use force. As Allison explained the Russian intervention in Syria, 'Putin's commitment to a global order which prizes the sovereignty of incumbent rulers remains to a large extent an external expression of his preoccupation with Russian domestic state order' (2013b: 818).

Finally, the war in Syria is another example of the complex interplay of cooperation and conflict in Russia's approach to the West. As Charap and Colton noted, rather than the intent to confront, the airstrikes represented a desire for inclusion, 'to break out of the diplomatic isolation and demonstrate that Russia could not be denied its rightful place at the high table of international politics' (2017: 163). On the one hand, Moscow's use of the military instrument as a means for enforcing cooperation predictably increased tensions and fuelled suspicions in the West of Russia's aggressive intentions. On the other hand, it made inevitable its inclusion as a central actor in future multilateral efforts to solve the crisis through diplomacy, like in the peace talks of the International Syria Support Group since autumn 2015, which involved Russia, the US, the EU, China, Iran and Syria, amongst others (Baunov 2015).

Conclusion

The wars in Ukraine and Syria have been interpreted as a 'paradigm shift' in foreign policy, where a 'revanchist' Kremlin, enabled by better military capabilities, is seeking to forcefully expand the country's influence in the CIS region and to confront the West in a bid for domination. The chapter has argued that a more contextualized analysis of these conflicts, taking into account Russian uses of military force since the early 1990s, is required for a better understanding of these events. It showed that there is little evidence of a fundamental change in Russian views on the utility of military force, or in its fundamental ambitions, both in the 'near abroad' and on a global level. Moscow has become more confident and assertive in using the military in pursuit of its national interest, not least because better capabilities have given it more opportunity to do so. However, today as has been the case in the past, the Kremlin perceives the military as a flexible tool of foreign policy and not only as an instrument to fight offensive wars.

Russian uses of military force in the CIS region have never been driven by the desire to materially recreate the Soviet Union. Moscow has used the military in this region for conflict resolution, to secure itself and other states against a variety of security threats, to protect its strategic and material interests, and to ensure the integrity of what it has consistently portrayed as its legitimate 'sphere of influence'. Russian military operations during the 1990s were couched in the language of peacekeeping. Even if these conflicts were dominated by Moscow, both by necessity and by design, attempts were made to gain legitimacy through a degree of multilateral cooperation. This approach was seen as both desirable and sufficient at the time, because Moscow worked on the assumption that its pre-eminent status in the region would be accepted by both its neighbours and the international community. Russia's conduct in the 'near abroad' became more assertive when the foreign policies of some CIS states took a decisive turn to the West, a process which, in the eyes of the Kremlin, the latter actively encouraged. Fearing the loss of its 'sphere of influence', Moscow abandoned the character of benign security guarantor and demonstrated, both in Georgia in 2008 and Ukraine in 2014, that its status in the region was non-negotiable.

Russia's experience of using military force beyond the CIS region is limited. However, past and recent examples suggest that the Kremlin's view on the utility of military power on an international level is not determined by the desire for global domination. In the Balkans, Russian peacekeepers contributed to conflict resolution in a multilateral setting. This enabled closer cooperation with the West and also improved the security situation in the region. In Syria, strategic interests and the growing influence of terrorist groups as a threat to stability in the region and beyond informed Russia's decision to intervene. On the global level, a major value of military power in the eyes of the Kremlin is its utility in ensuring the country's status as a great power. In Kosovo, the lack of such power resulted in humiliation and set in motion the process of military modernization. In Syria, a revived military ensured that Russia could not be sidelined in discussions and its views as a global power had to be taken into account.

Although Russia's military revival is unlikely to lead to further territorial expansion or an aggressive bid for global domination, the Kremlin's new confidence and assertiveness poses serious challenges to its neighbours and to the West. Moscow's preparedness to protect what it sees as its 'sphere of influence', if required by military force, is a threat to the sovereignty of the states in this region. It restricts their ability to pursue an independent foreign policy, and in particular their decision to join NATO or the European Union. With regard to the West, the danger of spiralling tensions and escalation cannot be dismissed. As the chapter showed, however, Russia is not an essentialist actor. Its foreign policy is not predetermined, but the result of interaction and, especially in the case of the West, of a mutual misunderstanding of intentions (Tsygankov 2016: 295). This does not mean that the West or anybody else is to blame for the Kremlin's actions. However, it shows that the nature of reactions and responses to Russia is likely to influence the course of future events.

Chapter 5

Russian military thinking and 'hybrid warfare'

The annexation of Crimea prompted a debate in the West about Russia's military revival and 'new military prowess' (Gordon 2014). As discussed in previous chapters, the operation revealed that the Russian armed forces had strengthened their capabilities in several areas, including 'enhanced deployability (tactical and strategic airlift), a relatively high level of training, and professional forces' (Reisinger and Golts 2014: 10). The air campaign over Syria, moreover, demonstrated that Russia now had the capability to engage in limited out-of-area operations beyond the area of the former Soviet Union for the first time (Gorenburg 2016). The perception of Russia's 'new military prowess', however, has been based not only on the augmentation of conventional capabilities. Structural and technological capabilities alone do not make a military powerful. As the history of warfare has shown time and time again, superiority in terms of equipment and manpower does not guarantee victory in war. Such superiority is an advantage and an important asset. However, military success also requires doctrinal and strategic thinking that is able to translate available assets into a tool relevant for the achievement of objectives in different and specific conflict scenarios.

What caught the attention of observers of the Crimea operation, above all, was the heightened sophistication in strategic and operational planning and conceptual development, denoted by the deployment of what has been described since then as 'hybrid' (Reisinger and Golts 2014), 'asymmetric' (Holmes 2014), 'ambiguous' (House of Commons Defence Committee 2014b), 'non-linear' (Galeotti 2014a), or 'next-generation' (Weitz 2014) warfare. It is the development of such 'new' approaches to warfare that is now widely seen as the most immediate potential threat to Russia's neighbours and to the West (Renz 2016b). This chapter discusses the concept of Russian 'hybrid warfare' in the context of developments in thinking on war and the utility of military force throughout the post-Soviet years. It will show that although there have been notable developments in Russian military thinking and doctrine in recent years, the concept of 'hybrid warfare' neither originates in Russian thinking, nor does it offer an adequate description of contemporary Russian strategy. The success of Russia's approach in Crimea should not be overstated. The operation showed that Russian military thinking has not been as stuck in Cold War concepts on conventional war-fighting as often presumed. However, it does not mean that Russian strategists have found the key to military success or a new war-winning formula that have put them ahead of the West in this respect.

Developments in Russian military thinking

Throughout much of the 1990s and 2000s, it was widely assumed that the Russian military could never modernize, because of the inability of its conservative leadership to move on from Cold War thinking on war and conventional, inter-state warfare with large armies (Renz 2014: 64). As Dmitri Trenin and Aleksey Malashenko argued, the lack of innovation in strategy and doctrinal adjustments during the first decade of the post-Soviet era was as much to blame for the failure of military reforms as the lack of funding. It is true, as the same authors noted, that when the Russian armed forces were established in 1992 they 'adopted de facto the last version of the Soviet military doctrine' (2004: 103–4). The Russian military's performance in the peacekeeping operations of the early 1990s, as discussed in the previous chapter, demonstrated that the lack of guidelines and doctrine pertaining to this kind of conflict caused significant problems. With the exception of thirty-six UN military observers sent to the Middle East in 1973, the Soviet Union had not participated in international peacekeeping operations until the late Gorbachev years, seeing them as 'first-world domination of less-developed states' (Yermolaev 2000). Unable, therefore, to draw on much existing experience in this area, Russian operations in Tajikistan, Moldova and Georgia were highly problematic, because they lacked, as Allison observed, 'commitment to the traditional principles of international peacekeeping', which they had never internalized (2001: 446).

The first Chechen War starting in 1994 also demonstrated that the lack of guidelines on how to deal with operations at the lower end of the conflict spectrum was a problem. Rather than basing the operation on an approach tailored to the circumstances of this insurgency, the military leadership 'seemed to react on instinct and poor intelligence, lashing out with bare hands, rather than a mailed fist', as Quentin Hodgson has put it. In what can be considered to be a line 'in the annals of gross misstatements', then-Defence Minister Grachev boasted at the outset of the war that the Russian armed forces would take Grozny in two hours, evidently assuming that numerical and technological superiority over the opponent was enough to guarantee swift victory (2003: 68).

Adjusting guidelines and doctrine to improve the Russian armed forces' preparedness to deal with scenarios other than large-scale, conventional inter-state wars has been an important aspect of the

2008 modernization programme. As the former Chief of General Staff Nikolai Makarov asserted, 'our military theory is outdated, since the 1980s the West has transformed its military capacities to fight wars of the future, but we have not done the same' (quoted in Bukkvoll 2011: 701). According to Makarov, efforts to deal with this problem included 'the reworking of all guidance documents, instructions, regulations and teaching aids'. However, by 2011 he noted that results in this area were not yet satisfactory, because they were still too much 'geared towards past wars' (quoted in Nichol 2011).

Discussions by analysts both in Russia and in the West of shortcomings in the adjustment of Russian military thinking during the 1990s and 2000s were often based on a simple dichotomy: 'outdated' Cold War thinking on the fighting of 'traditional' inter-state warfare that Russia had to abandon on the one hand, and innovative thinking on the fighting of 'new' wars and insurgencies it ought to move towards on the other. The idea that Russian military thinking until Crimea was dominated almost exclusively by conservatism and absence of future vision on war and war-fighting is not an adequate assessment. Contemporary Russian military thinking builds on a rich history, as this was a field of study where Soviet theorists produced influential and innovative work of international standing (Bukkvoll 2011: 683).

As Dima Adamsky noted in his extensive study of the culture of military innovation in various countries, the importance of traditional battle fought by large conscript armies and operational art relying on mass and moral superiority is but one important strand in Russia's military tradition (2010: 42–3). At the same time, Soviet military thinking stood out for its strength in 'theorizing innovative concepts', forward-looking 'outside the box' thinking, and the ability to create 'innovative and creative visions of ways to achieve victory' in wars of the future (2010: 52–3). For example, the idea of the so-called Revolution in Military Affairs (RMA), discussed in more detail below, which transformed US and Western views on the fighting and winning of future wars in the aftermath of the 1991 Gulf War, has its intellectual roots in Soviet military thinking on the 'Military-Technical Revolution' dating back to the 1970s. Analysing the likely long-term consequences of advanced technology for future warfare, Soviet theorists were ahead of their Western counterparts, as Adamsky noted, 'by almost a decade in its realization and in elaboration of the revolutionary essence embodied in US and North Atlantic Treaty Organization (NATO) military-technological shifts' (2008: 258). Russian thinking on asymmetric and 'indirect' approaches to warfare, which some observers saw as an innovation in Crimea, is also deeply rooted in Russian military tradition. In Adamsky's words, 'cunning, indirectness, operational ingenuity, and addressing weaknesses and avoiding strengths . . . have been, in the Tsarist, Soviet and Russian Federation traditions, one of the central components of military art' (2015: 25).

Innovative military thinking in Russia did not disappear with the collapse of the Soviet Union, and different strands of thought on how war was changing and should be fought now or in the future continued to develop. Tor Bukkvoll has usefully divided the major strands in contemporary Russian military thinking into three broad schools of thought, which he termed the 'traditionalists', the 'revolutionaries' and the 'modernists' (2011). The 'traditionalists', including the prominent strategist Makhmud Gareev, are closest to what one could call 'Cold War conservatism', in that they emphasize the importance of mass and traditional battle over technology. In their view, the potential threat of inter-state warfare with other major powers, such as the US/NATO and China, has been and still is Russia's primary security concern. Skills for fighting small wars and insurgencies are important, but secondary, and this should be reflected in the country's force structure and military posture.

The 'revolutionaries', most prominently Vladimir Slipchenko, are the heirs to Soviet thinking on the Military-Technical Revolution and their ideas on technology's transformational impact on future warfare. In their view, advanced technology will make the need for large standing militaries, and even the traditional division of army, navy and air force, redundant. Future wars will be 'contactless' and fought with stand-off precision weapons. In the 'revolutionaries' view, it will no longer be necessary to configure forces in correspondence to specific threats, as the technologically superior side will gain the upper hand against any enemy. Slipchenko's concept of contactless, 'sixth-generation warfare' is similar to what in the West is termed 'network-centric warfare', a central idea in the RMA. This is not surprising, because the turning point in military affairs, for Slipchenko and other 'revolutionaries', was and is the US coalition's victory in the 1991 Gulf War. Russian thinking on 'sixth-generation' warfare dates back to the early 1990s and it was also then when it caught the attention of Western defence establishments as a notable development. As noted in chapter 2, its central ideas were already reflected in the 1993 military doctrine's ambitious, yet at the time unachievable plans for developments in the Russian armed forces' conventional capabilities (FitzGerald 1994).

Finally, those whom Bukkvoll (2011: 697–701) labelled the 'modernists' are a less homogeneous group that includes notable thinkers like Aleksei Arbatov and Vitalii Shlykov. Broadly speaking, they present a middle ground, where desirable levels of manpower and technology are not considered in the abstract, but in relation to the country's current demographic and financial possibilities, as well as security requirements. In their view, trouble spots on the country's peripheries, such as the North Caucasus and Central Asia, are the most immediate security concern. These cannot be dealt with satisfactorily by a mass mobilization military, making improvements in rapid reaction a priority. Advances in technology are not seen as a panacea, but considered important for the maintenance of conventional war-fighting capabilities. The 'modernists' views on potential threats from NATO/the West and state actors, such as China, differ, with thinkers affording them various degrees of importance.

The existence of widely diverging views on the changing character of war and the utility of military force in military thinking are neither surprising, nor unique to Russia. When the Cold War ended, Russia, like many other states, was faced with the dilemma of being left with a military configured for a threat environment that no longer existed. It was uncertain what challenges the new international situation would bring in the immediate and longer-term future and what kind of armed forces and doctrines the country needed to deal with them. Ambiguities in Russian thinking on the merits of manpower versus technology, as well as divergences in views on what kind of conflict would dominate in the future, are also reflected in similar debates in the West (Renz 2014: 70-1). The question on how best to make a country's military suitable to the current and ever-changing security environment continues to be a matter of controversy not only in Russia, and, as long as chance and uncertainty exist, it can never be answered conclusively.

Developments in Russian military thinking are important, because many of the strategists driving these debates are close to the political and military leadership and their ideas inevitably shape military practice. However, the extent to which specific ideas and strands of thinking translate into changes to official military doctrine (discussed further below) is difficult to discern and also likely to change over time. As Bukkvoll noted, during the initial phases of the 2008 modernization programme, which was initiated in the aftermath of the war in Georgia, reforms were driven by the perceived need to improve capabilities required to deal with local conflicts and lower-intensity missions. The influence of 'modernists' calling for improvements in mobility and rapid reaction, therefore, seemed to be the most significant at this point (Bukkvoll 2011: 701). The 'modernists' perceived need to balance defence requirements with financial possibilities was also echoed by Putin in his early years as president, as noted in chapter 2. Following rising international tensions in the aftermath of the annexation of Crimea, however, priorities changed, as Bukkvoll observed in a follow-up work, because the West and NATO had 're-entered the stage as the main threat to Russian security'. This resulted in a partial reversal of the move from divisions to brigades on Russia's Western flank, indicating a reorientation back towards the 'traditionalist' camp (Bukkvoll 2016b).

The difficulty of identifying the exact influence of various strands of thinking prevalent in the Russian strategic community on military policy is salient for contextualizing the continuing debate on Russian 'hybrid warfare' capabilities. Following the annexation of Crimea, a number of Western analysts tried to trace back the origins of Russia's perceived 'hybrid warfare' approach to the writings of specific military thinkers. In particular, a now well-known article authored in 2013 by the Chief of General Staff, Valerii Gerasimov, caught the attention of various observers (Gerasimov 2013). Although neither Ukraine nor 'hybrid warfare' were mentioned in the article, it became known as the 'Gerasimov doctrine' and Gerasimov himself as 'the face of the hybrid approach' (Snegovaya 2015). As various experts specializing in Russian strategic thought later noted, the identification of Gerasimov's article as the origin of Russian 'hybrid warfare' thinking is selective and ignorant of how it fits into wider developments in Russian military thinking. Within this context, Gerasimov's ideas are not as 'new' as often asserted (Persson 2017: 2). As Charles Bartles showed, rather than revealing a 'new' way of war, Gerasimov outlined his views on the evolving trends in Western and US approaches to warfare, which he traces back to the 1991 Gulf War. Outlining his view of future military operations, he envisioned 'less large-scale warfare; increased used of networked command-and-control systems, robotics and high-precision weaponry' (Bartles 2016: 36). As such, the article was a clear continuation of 'revolutionary' thinking on the future of warfare and sixth-generation war.

A similar assessment has been made of a second piece hailed as a blueprint for the Crimea operation – an article entitled 'The nature and content of a new-generation war', written by two officers of the Russian General Staff Academy, S.G. Chekinov and S.A. Bogdanov in 2013 (e.g., Perry 2015). Defining the twenty-first century as the 'age of high-tech wars', their thoughts, according to Timothy Thomas, were by no means an innovation. Instead, he noted that 'Slipchenko's work [on sixth-generation warfare] may be considered as a (or the!) most important source behind much of the new-generation warfare thinking and postulating of Bogdanov and Chekinov' (2016: 555). This is not to say that there were no notable new ideas in these articles to justify the attention they received. Both articles' authors highlighted the growing importance of non-military tools (such as sanctions, political pressure and information campaigns) in contemporary warfare which, in their view, now exceeded the potency of physical military force. This did not indicate, however, that they no longer saw conventional capabilities as essential or that Russian strategists intended 'to remove, as far as possible, displays of "hard military power" from modern warfare with "war" becoming something fought at "arm's length" without the need to engage with adversary's forces', as Rod Thornton has cautioned (2015: 44). As Frederik Westerlund and Johan Norberg pointed out, even during the Crimea operation, the display of 'hard military power' in the form of a 150,000-man snap exercise near the area of operation was part of the mix to ensure Russia's potential to dominate escalation and to deter interference or response from either Ukraine or NATO (2016: 591).

Russian military thinkers advanced a variety of ideas on future war and warfare throughout the 1990s, and these ideas have influenced military policy to various degrees. The recent focus in the West on Russian 'hybrid warfare' reflects a one-sided view of these developments. Although the growing emphasis on non-military tools is an important change in Russian strategic thought, it is important not to take this change out of context, as this could lead to a skewed understanding of Russian military capabilities and ambitions. As Dave Johnson has aptly observed, and as is also evident in the direction of military modernization discussed in chapter 2, Russia continues to rely 'on at least the leveraging, and potential employment, of full-spectrum conventional, unconventional and nuclear capabilities' (2015: 2). Or as Bartles concluded, 'Russia is experimenting with some rather unconventional means to counter hostile indirect and asymmetric methods, but Russia also sees conventional military forces as being of the utmost importance' (2016: 36). Russian military thinking is developing as lessons are learned from past operations and adjustments are made to the changing security environment. 'Hybrid warfare', which is discussed as a concept in more detail below, does not adequately reflect the essence of contemporary Russian military thinking. As Thomas warned in 2016, 'it is doubtful that this progression will stop here, since new forms and methods are under development every month from lessons learned in Syria, in Ukraine, and in dealing with the Arctic. Putting Russia's military thought in a specific box . . . is a mistake, as it is evolving and changing over time' (2016: 555).

Developments in Russian official military doctrine 1993–2014

Developments in official Russian military doctrine, a document that is updated and published every few years, are an indicator of continuity and change in Moscow's views on war and strategy. Although Russian military doctrine focuses on the political-strategic level of war and is not a set of guidelines for tactical and operational military action as in the West, variations in threat perceptions and elaborations of possible responses serve as a useful gauge of strategic views and priorities (De Haas 2011: 3). As demonstrated in the overview of Russian military doctrines since 1993 below, the need to maintain and develop strong, conventional war-fighting capabilities has been emphasized since the early 1990s. Different versions of the doctrine also show a growing recognition that the development of skills required for dealing with lower-intensity missions are important. Although the latest military doctrine adopted in 2014 included the threat of outside information influence over the population as a military threat and thus reflected the increasing importance attached to non-military tools, there is little evidence in official military doctrine that 'hybrid warfare' has become the focal point of Russian strategy.

Russian military doctrines were updated in 1993, 2000, 2010 and 2014. The doctrines' respective focus was adjusted to reflect changes in the international security environment and in response to events perceived as significant to Russia's national security and interests. However, all doctrines show a substantial degree of continuity in threat perceptions and strategic priorities. The 1993 document was the first military doctrine adopted by the Russian Federation and supplanted the last Soviet military doctrine, on the basis of which the country's armed forces still officially operated. As it was issued at a time of relatively low tension with the West in the immediate post-Cold War period, it reflected a positive view of international relations at the time. This quickly changed and grew increasingly pessimistic in subsequent versions. The threat of internal armed conflict was prioritized as a military danger in the 1993 doctrine, because secessionist tendencies were the most immediate challenge to Russian statehood and the threat of inter-state warfare had diminished. All subsequent doctrines demonstrated a growing awareness of changing international security priorities that also pertained to Russia. 'New' challenges to Russian security and stability in the form of extremism, ethnic strife and religiously motivated terrorism, both within the country and in its neighbourhood or perceived sphere of influence, took an increasingly important place.

At the same time, 'traditional' threat perceptions arising from growing tensions with the West and NATO eastward enlargement in particular have been a constant feature in all versions of the military doctrine. The encroachment of NATO into what Russia perceives as its legitimate sphere of influence in the former Soviet region has been a particular concern from the early post-Soviet years. The 1993 doctrine already referred to 'the expansion of military blocs and alliances to the detriment of Russian security interests' as a potential military danger (The Basic Provisions of the Military Doctrine of the Russian Federation 1993: section 2, paragraph 1). This concern was reiterated in the 2000 doctrine (The Military Doctrine of the Russian Federation 2000: section 1, paragraph 5). The 2010 and 2014 doctrines, in a reflection of deteriorating relations with the West following the Kosovo War and the annexation of Crimea, explicitly named NATO expansion and the movement of NATO military infrastructure closer to Russia's border as the top main external military danger to Russian security (The Military Doctrine of the Russian Federation 2010: section

2, paragraph 8; The Military Doctrine of the Russian Federation 2014: section 2, paragraph 12).

The importance of Russia's strategic nuclear deterrent was also emphasized in all doctrines. The 2000 doctrine introduced the provision that nuclear weapons could be used in response to a conventional aggression, 'in situations critical to the national security of the Russian Federation' (section 1, paragraph 8). This point was reiterated in both subsequent versions, which specified that the country reserved the right to use nuclear weapons if conventional aggression by an external actor threatened 'the very existence of the state' (2010, section 2, paragraph 16; 2014, section 3, paragraph 27). All doctrines have consistently emphasized the desire to strengthen the armed forces' conventional capabilities and non-nuclear deterrence. The ongoing inclusion of the provision that nuclear weapons can be used against conventional aggression in certain circumstances in the 2014 doctrine indicates that serious doubts about Russia's comparative conventional military power persist.

The concurrent emphasis in Russian military doctrine on 'small wars' and 'traditional' inter-state warfare was often regarded by observers as an inconsistency in military thinking that was partially to blame for the failure of reforms up until 2008. In the view of authors critical of this perceived inconsistency, successful reforms and modernization required the unambiguous choice of the former over the latter. As Margarete Klein argued, in the post-Cold War security environment, Russia needed troops that were 'more flexible and professional and, therefore, combat-ready for scenarios like local conflicts and asymmetrical warfare'. This could not be achieved, though, as long as Russia hung on to 'the old concept of a mass-mobilization army' (2012: 43). In fact, this perceived 'inconsistency' was not an oversight, but a reflection of the threat perceptions driving developments in Russian military doctrine. Unlike in the West, where defence establishments and strategists prioritized counterinsurgency doctrine and increasingly considered large-scale inter-state warfare a thing of the past, in Russia the 'traditional Western threat', as Trenin and Malashenko have put it, had returned to the centre of strategic thinking already as early as the mid-1990s (2004: 104). Small wars and insurgencies were never seen in Russia as the only scenarios its armed forces had to be prepared for.

In the run-up to the publication of the 2014 military doctrine, there were concerns that its focus would be revised significantly in view of growing tensions with the West. Some feared that it might introduce the concept of nuclear pre-emption or name specific countries as its enemies for the first time. In the event, this did not happen. The portrayal of NATO eastward enlargement as a major military threat had already been evident in the 2010 doctrine and was merely carried over. The 2014 version dropped its reference to cooperation with NATO as a means to reinforcing collective security, now only mentioning the alliance as a potential partner for 'equitable dialogue' (section 3, paragraph 21).

Having said this, one significant discontinuity and change can be observed in the 2014 version. For the first time the doctrine included in its section on domestic military dangers the notion of external threats to 'the information space and the internal sphere'. Specifically, it referred to the danger of 'the informational influence over the population . . . aimed at undermining spiritual and patriotic traditions'. The doctrine repeatedly affirmed the need to strengthen state policies aimed at countering such outside influence into Russia's domestic affairs (Sinovets and Renz 2015: 2). Concerns over the security implications of outside political influence and the country's 'information security' are not new to Russia. 'Information management' aimed at shielding the population from outside influences was, of course, central to the domestic politics of the Soviet Union. Even during the 1990s, at a time of relative media freedom in Russia, there were fears, as Thomas noted, that 'in an unstable public-political and socio-economic situation, the entire population could serve as the target of influence for an enemy campaign'. Management of information was seen as essential to the maintenance of stability in the country (1996: 31). As is well known, the control of information in Russia has become increasingly tight since Putin's rise to political power. Already in 2000 a document entitled the 'Information Security Doctrine of the Russian Federation' was adopted. This included subjects such as the moral content of the media, and clearly indicated the prioritization of the information sphere as a matter of national security (Bacon et al. 2006: 89–91). Until 2014, however, this issue was never explicitly addressed in the military doctrine.

The inclusion of outside information influence over the population in the 2014 military doctrine clearly confirms the Kremlin's increasing concern over internal order and regime stability as also discussed in chapters 3 and 4. This is where the influence of certain strategic thinkers on military policy is clearly evident. The above-mentioned emphasis by strategists like Gerasimov on the growing potency of non-military tools in contemporary war and conflict needs to be understood within this context. Rather than outlining a Russian approach or doctrine of 'hybrid warfare', a concept that is never mentioned in such a context either by Gerasimov or by other strategists, his discussion is concerned with the potency and danger of non-military tools, such as political/ information influences on the Russian population, pursued by the West. As Bartles stated, rather than outlining Russia's future approach to warfare, Gerasimov's article highlighted his view of 'the primary threats to Russian sovereignty as stemming from US-funded social and political movements, such as color revolutions, the Arab Spring, and the Maidan movement' (Bartles 2016: 36). This yet again confirms that Russian military power is as much about internal stability as it is about the actual fighting of wars with external enemies.

'Hybrid warfare' – what's in the label?

Some policy makers and analysts in the West interpreted the ease of Russian victory in Crimea as evidence of a new Russian approach to warfare that posed a serious threat to its neighbours and to the West. A UK House of Commons Defence Committee report in 2014 cautioned that the use of Russian 'asymmetric tactics (sometimes described as unconventional, ambiguous or non-linear warfare) . . . represents the most immediate threat to NATO neighbours and other NATO member states' (House of Commons Defence Committee 2014b). According to Thornton, this new approach appeared to be particularly difficult for Western powers to stand up against, as it allowed Russia to overcome shortcomings in its conventional military power and 'to negate the significant advantage held by the US and its NATO allies in terms of their conventional military force, mostly in the technological realm' (2015: 44). 'Hybrid warfare' does not originate in Russian military thinking. Although it has now also entered the parlance of strategic thinkers and military leaders in Russia, it is never referred to as a Russian approach, but, as discussed above, as a method used by the West against Russia (Thomas 2016: 557).

The 'hybrid warfare' concept originated in the US and was first conceptualized in depth during the mid-2000s by Frank Hoffman, a former officer in the US marines. In his view, the idea of 'hybrid warfare' usefully encapsulated important changes pertaining to the utility and application of military force in the twenty-first century and could explain in particular the success achieved by comparatively weak opponents against technologically advanced and superior militaries, such as the Taliban or Al Qaeda in Afghanistan and Iraq (2007). What makes an approach to warfare 'hybrid' is the coordinated use of tactics combining military force with non-physical or unconventional modes of warfare, such as terrorist acts, criminal disorder, the use of proxy forces, psychological and information operations. In the aftermath of Russia's swift and almost bloodless campaign resulting in the annexation of Crimea in 2014, the 'hybrid warfare' concept gained widespread popularity in the West, because it appeared to be an accurate description of Russia's approach in this operation, where non-military tools and the use of information played a central role.

One problem with this interpretation is that it is based on the assumption that there is such a thing as a war-winning strategy that will work irrespective of the circumstances specific to each war. As the history of strategic thought shows, 'hybrid warfare' is not the first concept that seemed to promise universal military success. Traditionally, any such concept failed to deliver decisive results in subsequent applications and it is doubtful whether Russia has finally found the elusive silver bullet. A prominent example of a presumed war-winning formula is the above-mentioned RMA. Although the intellectual roots of the concept originated in Soviet thinking on the implications of technological innovation on the nature of war, it became prominent in US strategic thought in the wake of the 1991 Gulf War (Adamsky 2008). Some strategists extrapolated from this example that the US had found an approach to war that would allow it to overcome the 'fog of war' in any future conflict. The air campaign over Kosovo, in the eyes of the concept's proponents, vindicated this idea (Adamsky 2010: 3). As is well known, the RMA did not turn out to be a panacea and US technological superiority did not deliver success in the subsequent conflicts in Afghanistan or in Iraq in 2003.

Although technological superiority undoubtedly presented the US coalition with a serious advantage during the first Gulf War, it is also clear that victory in this case was enabled by exceptionally favourable conditions for the US coalition. As Tim Benbow summed it up, these included a united international opinion against Iraq. The latter had invaded another sovereign state and thus presented a clearly defined strategic objective to the coalition. This was the liberation of Kuwait, which suited the conventional advantages of the coalition and could be achieved by conventional military means. The terrain and weather made the theatre of operations particularly conducive for the success-ful conduct of the air campaign and the coalition had a long build-up period to prepare for the operation. It also enjoyed host nation sup-port in the region and the opponent it faced was a regular military that operated with similar tactics, but was far inferior in terms of technology and manpower (Benbow 2004: 56–8). Less favourable circumstances meant that technological superiority did not translate into easy mili-tary victory in Afghanistan and Iraq in 2003. Mountainous terrain and a geopolitical location unsuitable for easy logistical support compli-cated the situation in Afghanistan. Patchy international support for the 2003 Iraq War was a problem. Arguably, however, the ambiguous and fluid strategic endpoints in both cases made the swift achieve-ment of objectives unlikely from the outset. Unlike the liberation of an occupied country as in 1991, the goals of fighting global terrorism, achieving regime change and effecting stabilization were not suited to the military advantages aiding US victory in 1991. Coalition forces got bogged down in protracted campaigns where, once the existing regime had been defeated, insurgent groups were able to ensure that a lasting US strategic victory could not be achieved. The chances of success of any war-fighting approach are never universal, but always depend on context and circumstances.

Developments in Western military thinking related to counterinsurgency doctrine are also relevant for the debate of 'hybrid warfare' as an assumed new Russian approach to war. Drawing on the historical experience of Britain, France and the United States, counterinsurgency doctrine was proposed as an antidote to the US coalition's military failures in Afghanistan and Iraq. Although there is no single template for how counterinsurgency operations should be approached, there is widespread agreement on the need to gain the support of the local population. This is essential to ensure their participation in the counterinsurgency efforts or at least to prevent them from supporting insurgent groups (Kieras 2016: 351).

Information and psychological operations, conducted by units trained specifically for this purpose or by special operations forces, are an important tool for influencing and gaining support from the local population. As Robert Egnell has put it, success in 'winning the hearts and minds' of the population is essential for successful counterinsurgency warfare, because 'the number of battlefield victories . . . matter little if the population thinks you are not winning, or visibly improving people's situation' (2010: 291). At the same time, although it is generally accepted that getting the local population 'on side' is an absolute requirement for defeating an insurgency, there is a 'staggering lack of empirical evidence to support' that it is routinely delivering expected outcomes (Egnell 2010: 292). Especially when it comes to information and psychological operations, the lack of understanding and insight into the historical, social, political and economic context of the target state makes it difficult to communicate the intended message. As Egnell concluded, 'insurgents have the inherent advantages of better cultural understanding and closer contacts with the local population.

They are therefore in a strong position to present alternative narratives to events, and even turning tactical losses into victories of perception' (2010: 291).

The examples of the RMA and of counterinsurgency warfare provide some useful context for the study of 'hybrid warfare' as a supposed war-winning Russian strategy. As the case of the RMA shows, drawing generalized conclusions from a single successful application of a war-fighting approach is problematic. The potency of Russian 'hybrid warfare' capabilities should not be taken for granted, because the Crimea operation to date is the only example of the successful application of this presumed approach. Moreover, the Crimea operation, which was very limited in scope and scale, does not provide sufficient empirical evidence to conclude that tactics successfully used in this case will deliver a similarly positive outcome in a different scenario. From this point of view, assessments of the Crimea operation as a new Russian mode of warfare, allowing the Kremlin to circumvent relative weaknesses in its conventional military power vis-à-vis the West, should be taken with a pinch of salt.

The relevance of the RMA and counterinsurgency discussion for the debate on Russian 'hybrid warfare' goes beyond this obvious point. As was the case for the US coalition in the 1991 Gulf War, Russia operated in Crimea under extremely favourable conditions that negated the need for overwhelming force and enabled the swift annexation of the peninsula within a matter of weeks (Norberg et al. 2014: 44–7). The fact that Russia already had a considerable military presence in Crimea eased the logistics of the operation and enabled the Kremlin to prepare for the operation without the need for conspicuous troop movements. As a result of the political turmoil in Ukraine leading up to the operation, the country's political and military leadership was weakened and unable to stage a coordinated response. This was exacerbated, as

Maxim Bugryi wrote, by the government's 'tenuous political control over the Crimean peninsula' (2014). It is also likely that Russia acted in the knowledge that it could operate with relative impunity, as outside military assistance to Ukraine was improbable. There was little danger the annexation of Crimea would lead to a direct military confrontation with NATO (Adamsky 2015: 38). This possibility would certainly have to be factored into Russian risk calculations if it resorted to direct aggression, even by 'hybrid' means, against one of the alliance's members.

Arguably the most important factor enabling the success of the 'hybrid' approach in Crimea was the strongly pro-Russian outlook of the majority ethnic Russian local population, making a protracted insurgency unlikely. This negated the need for the use of overwhelming force from the outset and meant that a combination of mostly non-military 'hybrid' tools was sufficient. In Crimea the usually difficult battle for 'hearts and minds' was won before it even had begun. Cultural misunderstandings were not an issue and there was a negligible danger of a popular uprising or significant support for potential insurgent forces. There was no need for Russian soldiers to engage in messy battles or risk civilian casualties and they were left to secure government buildings and Ukrainian military installations with next to no resistance. It is easy to see that such conditions would be difficult to replicate in any setting other than Crimea. Serious resistance leading to protracted violence and use of military force would be almost inevitable even in neighbouring states with significant Russian minorities. It is unlikely that a 'hybrid approach' as used in Crimea would have enabled easier victory for Russian forces in previous interventions, such as Chechnya and Georgia. Although in spite of all of its shortcomings the Russian military was vastly superior in terms of manpower and technology, Chechen fighters had sufficient forces and willpower to stage a sustained campaign of resistance and to inflict heavy losses on Russian troops. Improved capabilities to deal with lower-intensity missions would have meant that the wars in Chechnya could have been fought with less brute force and fewer casualties on both sides. However, given the Chechen side's willingness to resist, it would never have been a 'bloodless' victory. Moreover, winning the 'hearts and minds' of the Chechen population would still have been difficult.

Unlike in Crimea, where significant portions of the population were in favour of joining Russia, the Chechen wars were fought to quell the republic's desire for independence. Although the increasingly radical Wahhabist insurgents during the second Chechen War never enjoyed widespread popular support, historical antagonism and feelings of revenge meant that Russian troops were far from being welcomed with open arms (Kramer 2005: 215). No amount of 'hybrid warfare' would have changed that. It is unlikely that a 'hybrid' approach would have worked in the war in Georgia in 2008. This was fought as a conventional military campaign and combined arms operation. Such an approach was suitable to the war's objective, which was to expel Georgian forces from South Ossetia and to weaken the country militarily. In this conflict, Russia fought a conventional opponent using similar equipment and tactics, but the Russian military had the upper hand through its overwhelming physical superiority. Although the war revealed some operational weaknesses, including problems with command, control and communication, the aim of expelling Georgian forces was achieved within the matter of five days (Bukkvoll 2009: 61). It is unlikely that 'hybrid warfare' would have made a substantive difference to the outcome of the campaign.

The importance of gaining support by the target population is particularly salient for estimating information warfare capabilities, which are increasingly seen as a dangerous aspect of Russian 'hybrid warfare'. Former NATO SACEUR Philip Breedlove called the annexation of Crimea the 'most amazing information blitzkrieg . . . in the history of information warfare'. This sentiment was echoed by Thornton, who concluded that 'the major threat to Western interests anywhere in the world is . . . the threat posed by information warfare such as that recently conducted by Russia. It has achieved clear results, and this success can be repeated' (2015: 40, 45). It is beyond doubt that Russia has stepped up its efforts to improve its information warfare capabilities, both in the technological (electronic warfare and cyber operations) and psychological (information/disinformation operations and deception) realms. It is also clear that the Kremlin is using information in various guises (social media and 'twitter trolls', state-owned media outlets, such as Russia Today and Sputnik) as a foreign policy tool to seek political influence abroad and that this poses challenges to its neighbours and to the West (Kofman and Rojansky 2015). With regard to the latter, it is at least debatable whether this type of use of information should be discussed under the umbrella of 'hybrid warfare' at all or whether this represents, as Samuel Charap wrote, a 'dangerous misuse of the word "war"' (2015/16: 52).

The generalizability of conclusions from the success of Russian information operations in Crimea *per se* should not be overstated. As the experience of many militaries using information operations during war has shown, they rarely achieve conclusive results and often fall short of expectations (Jackson 2016). In Crimea, Russian information operations were effortless as their primary target was receptive to the Kremlin's narrative of events from the outset. The success of Russian information operations beyond Crimea cannot be taken for granted, however. For example, a large-scale study assessing the influence of official Russian narratives and disinformation during the Crimea crisis in eight European countries found that this influence was 'largely limited' (Pynnöniemi and Rácz 2016: 312). Moreover, Mark Galeotti concluded that the major 'success' of Russian information warfare since its intervention in Ukraine seemed to have been the growth of negative views of Russia across Europe from 54 to 74 per cent within the space of a year (2015).

Favourable circumstances in Crimea meant that Russian 'hybrid warfare' in this case yielded impressive results. Russia's approach in Crimea worked, because it applied suitable means to a specific end. It made 'concrete a set of objectives through the application of force in a particular case', which is evidence of good strategy (Strachan 2013: 12). It is not, however, evidence of the invention of a new, war-winning formula that can easily be replicated in a different, less favourable scenario.

What 'hybrid warfare' can and cannot explain

The concept of 'hybrid warfare' has been useful inasmuch as it highlighted a number of new capabilities demonstrated in Crimea and it showed that the Russian military was not as stuck in Cold War thinking as previously assumed. As Antulio J. Echevarria noted, new strategic concepts are useful in this respect, as they can help draw the attention of policy makers to emerging security challenges. As he also pointed out, however, there is a tendency for concepts such as 'hybrid warfare' to turn into a dogma not supported by strategic analysis that can hinder, rather than aid, decision making and strategic planning in the long term (2015: 16). There is evidence that this is already happening in some Western analyses, where almost every Russian foreign policy move is now interpreted as a part of a 'hybrid campaign'. This is decreasing the explanatory power of an already vague concept. As Michael Kofman has put it, 'if you torture hybrid warfare long enough it will tell you anything . . . The term now covers every type of discernible Russian activity, from propaganda to conventional warfare, and most that exists in between. What exactly does Russian hybrid warfare do, and how does it work? The short answer in the Russia-watcher community is everything' (Kofman 2016).

The concept of 'hybrid warfare' first gained traction as it seemed to offer a good explanation for how Russia achieved its swift and almost bloodless victory in the Crimea operation. If Crimea is considered the 'gold standard' of Russian 'hybrid warfare', its relevance for the analysis of other Russian military interventions is certainly questionable. It is widely held that the ongoing fighting in eastern Ukraine is the extension of an ongoing Russian 'hybrid war' against Ukraine (e.g., Iancu et al. 2015). In fact, apart from Russia's involvement, there is very little similarity, from a strategic point of view, between the Crimea operation and the fighting in Donbas. As Sibylle Scheipers pointed out, Russia used a war-shortening approach in Crimea, where surprise, tempo and superior information was used to conclude the conflict before major battle could even commence. In contrast, in Donbas Russia pursued an indirect approach of the war-lengthening variety, drawing the civilian population into the conflict and relying on a mix of auxiliary fighters and Russian military personnel (Scheipers 2016). Unlike Crimea, the war in Donbas has been far from swift or bloodless. Whether intended or not, the conflict continued into 2016, resulting in over one million internally displaced persons. The death toll approached 10,000 casualties, most of them civilians, by the summer of that year. Even if long-term destabilization of the region was the desired objective, there have been doubts about the degree of Moscow's control over the fighting. Moreover, there have been costly unintended consequences of the 'hybrid approach' pursued in Donbas, such as the downing of the Malaysian airliner MH17 (Galeotti 2014b). Although some features of the war in eastern Ukraine can be described as 'hybrid' inasmuch as it involves a mix of conventional and 'irregular' tactics, this certainly does not support the idea that such an approach is universally successful. In fact, as Paul J. Saunders noted, 'what worked in Crimea demonstrably failed in eastern Ukraine' (2015).

Russia's involvement and intervention in the Syrian civil war has also been described as 'hybrid warfare' (Cordesman 2015). This is particularly puzzling, as it is hard to see significant similarities between the approaches pursued in both conflicts. In contrast to the 'contactless' war in Crimea, Russia's Syria intervention in terms of tactics and technology took the form of a conventional air campaign not dissimilar to Western air-only operations pursued over the past two decades. Russian air operations in Syria have been far from 'bloodless' and showed little concern for civilian casualties. According to Ari Heistein and Vera Michlin-Shapir, 'although the future of the war in Syria is uncertain, what remains clear is that Russia is fighting a hybrid war, combining its military, diplomatic and media capabilities to achieve its goals using limited armed engagement' (2016). This stretches the concept too far and what the authors describe as 'hybrid warfare' is, in fact, basic grand strategy – the level of war where all instruments of power at a state's disposal are routinely combined towards the achievement of political objectives (Liddell Hart 1967: 335–6). A combination of diplomacy, information, intelligence and economic tools have traditionally been used in most wars and as such this is not an expression of 'hybridity'. The success of Russian information aimed at influencing international opinion on its involvement in the Syria conflict is also questionable. It is one thing to observe that Russia is using a new 'technique of narrative construction and control through its international media outlets' in order to build a 'triumphalist narrative' depicting 'Russian bravery and resolve', as Heistein and Michlin-Shapir (2016) had put it. It is quite another thing for such a narrative to have traction amongst audiences outside of Russia and in the West. The widespread coverage in the Western media of civilian atrocities caused by Russian airstrikes and accusations of Russian war crimes at the highest level, including in the UN, imply that the international influence of the official Russian narratives is at best severely limited, if not counterproductive.

Finally, the tendency to describe Russian actions not linked to specific military operations as 'hybrid warfare' is particularly problematic. What is essentially a military concept describing a tactical approach to war has morphed into a quasi-theory of Russian foreign policy which, in the eyes of some observers, explains almost every (perceived) Russian move: Russian internet trolls (Spruds et al. 2016), official political statements (Amann et al. 2016), refugees crossing the Russian border (Higgins 2016), even football hooligans (Jones and Smith 2016), have all been described as part of an elaborate plan of Russian 'hybrid warfare' against the West. Such a broad use of the concept stretches it to the extent that it has become meaningless. 'Hybrid warfare' describes a mix of military and non-military means and there is nothing 'hybrid' about the use of information or political statements in isolation. The idea that Russia is conducting a 'hybrid war' against the West grossly oversimplifies Russian foreign policy which, as Trenin showed, is determined by various geopolitical concerns and policy drivers (2016a). A broad and all-encompassing label, such as 'hybrid warfare' might be appealing, but it is unable to capture the nuances of Russian foreign, security and defence policy and ultimately will make it harder to identify realistic responses vis-à-vis Russia in the long term.

Conclusion

Innovation and revival in Russian military thinking, based on the operations in Crimea and Syria, should not be exaggerated. Developments in the country's strategy and doctrine show both continuity and change and cannot be summed up as a simple process of transition from the focus on 'traditional' wars to 'new' wars or 'hybrid warfare'. During the years leading up to the annexation of Crimea, the extent to which Russian military thinking continued to be based on Cold War assumptions about war and conflict was overstated. Although it is true that when the Russian armed forces were first created in 1992 they initially continued operating under Soviet military doctrine, Russian military and doctrinal thinking reflected the challenges arising from the changing security environment and their strategic implications. External state actors and NATO continued to occupy an important place in Russian threat perceptions for much of the post-Soviet era. However, this was not the result of the inability of conservative military thinkers to move on from the Cold War, but reflected the country's strategic priorities and threat perceptions, which differed from those of the West.

As such, the assumption that the Russian military could only become truly 'modern' if it followed the Western example in adjusting its structure and doctrine was a fallacy, because it did not take into account the idiosyncrasies specific to the country. As it is the case with Russia's conventional military capabilities, overestimation of its doctrinal and strategic prowess in the aftermath of Crimea now appears to be the greater danger. There is little evidence to support the idea that Russian innovations in war-fighting have put it ahead of the West in this respect or that such innovations are enough to make up for shortcomings in its conventional capabilities. The idea that Russian military thinking took an almost complete about-turn and changed

ιscence to 'hybrid warfare' wizardry in the matter
ιnrealistic. The operations in Crimea showed
ιͻ the wars in Chechnya and Georgia, Russia has
ιved its ability to fine-tune and adjust military tactics to the
ιcumstances of operations of various scope and intensity. However, it has not found a new key to universal military success in the form of 'hybrid warfare'. Operational success in Crimea was enabled above all by a combination of unique and favourable circumstances. The success of Russian military strategy will continue to be subject to the effects of chance, uncertainty and the 'fog of war'.

Conclusion

Russia experienced a military revival in recent years. From 2000 onwards, and especially since the military modernization plans announced in 2008, its conventional military power has vastly improved. During the 1990s, the degradation of military capabilities contributed to the country's international image as a power in decline. Today, Russia is well on the way to overcoming this problem. The implications of the military revival go beyond these improvements in capabilities and image. Russia's confidence in using the military as a tool of foreign policy has also grown. The annexation of Crimea and the airstrikes in Syria demonstrated that Moscow today is not only able, but also willing, to pursue what it sees as its national interests, even in the face of strong international condemnation and on a global level. Russia is yet again a power to be reckoned with.

This book argued that, although improvements in Russian military power and the Kremlin's growing confidence to use this instrument are certainly significant, the implications of these developments are not as straightforward as often assumed. Russia's relative military power compared to the West remains limited and there is also no evidence of a fundamental turnaround in Moscow's views on the utility of force. It cannot simply be assumed that the military revival signifies Russia's desire to threaten its neighbours and the West in a bid for domination. This reflects an incomplete understanding of why Moscow sees powerful armed forces as important. The military revival did not occur in a vacuum. It can only be understood within the context of relevant developments in the country's history. The book's chapters studied the role of the military in foreign policy in the past and today, reforms of the armed forces since the early 1990s, the meaning of the force structures as an aspect of Russian military power, Moscow's use of military force since the end of the Cold War, and changes in the country's military thinking. As such, they provided the context required for a more informed assessment of recent events.

The following sections return to the three major assumptions underlying Western reactions to Russia's military revival as outlined in the introduction. Addressing each assumption in turn, they will summarize the book's main arguments about why these assumptions are problematic. In the final section, possible options for Russia's neighbours and the West in dealing with an increasingly assertive Russia will be addressed.

The timing of Russia's military revival

Russia's military revival is not the result of a sudden turnaround in Russian foreign policy. An area where there has been a significant degree of continuity, rather than change, is Russia's self-perception as a great power and its desire to be granted this status by the international community. The need to maintain an internationally competitive military, as an essential characteristic of a great power, has always been central to this self-perception. This did not change when the Soviet Union collapsed in 1991. As the country's conventional military capabilities degraded throughout the 1990s, it became ever clearer to the political leadership that Russia's status as a global military power could not be upheld with a strong nuclear deterrent alone. Other global powers continued to develop their conventional capabilities and the utility of nuclear weapons for the protection and pursuit of Russian interests in the changing international environment was limited. Ambitions to revive Russia's military power and to remain competitive with the West in this respect date back to the early 1990s and beyond. Owing to a complex confluence of political, societal and financial factors, these ambitions did not start to turn into reality until the turn of the millennium and the implementation of a systematic programme of military modernization in 2008.

Whilst Russia's self-perception as a great power and its views on the importance of being able to project global military power have not changed substantially, international views of Russia as a military actor have certainly experienced a revival. As the country's conventional military capabilities degraded throughout the 1990s, so did its ability to project such power beyond the immediate neighbourhood. Russia's relevance as a global military actor, with the exception of its continuing status as a nuclear power, was increasingly questioned. Following the annexation of Crimea and intervention in Syria, views on Moscow's status in the world have changed dramatically. The international community is again regarding Russia as a serious competitor on the global stage, thus, at least partially, granting it the status recognition it has been seeking ever since the fall of the Soviet Union. From this point of view, Moscow has to some extent achieved one of its long-standing objectives. International recognition as a great power is key to Russia's national interests and has been a central aim of its foreign policy throughout history. Now that Russia is yet again seen as a global military power, it has gained such recognition, but this is not based on respect.

Assertive uses of military power in the CIS region have served to reassert Russia's position as the dominant actor there. At the same time, aggressive actions, such as the war with Georgia in 2008 and with Ukraine in 2014, have decreased the country's prospects of being the hegemon in what it sees as its 'sphere of influence'. These actions have exacerbated fears by neighbouring states about Russian dominance that date back to the early 1990s. As a result, many CIS states might even be more determined to pursue closer relations with actors other than Russia. As such, the Kremlin's quest to gain great power recognition through the display of military power has been a double-edged sword. As lessons from the past reveal, pursuing status based on military might alone, whilst remaining weak in other areas and especially in the economic realm, has not been sustainable and never led to lasting results.

The reasons for Russia's military revival

It cannot be taken for granted that Russia's military revival signals a qualitative change in the Kremlin's intentions to use force in an expansionist fashion or a bid for domination. Better capabilities create more opportunities for the use of force, but they do not necessarily increase the willingness to do so. The assumption that Russia has revived its military in order to prepare for more offensive action reflects a reductionist view on the utility of military power and the reasons why most states, including Russia, maintain strong armed forces. Developments in military doctrine show that the reasons for the military revival go beyond the desire to fight wars and defeat opponents. The requirement to defend Russian sovereignty and territorial integrity remains the number one task of the Russian armed forces. The need to have capable and sizeable armed forces to protect the country against external threats has been particularly acute for Russia throughout history, owing to its size and complex geopolitical position. During the Cold War, threat perceptions in the Soviet Union, as was the case for most countries in the West, focused on the potential of East–West confrontation. Soviet military doctrine therefore emphasized the need for preparedness to engage in large-scale conventional warfare. There is a strong degree of continuity in Russian military doctrine today with regard to the importance attached to this type of war-fighting and threat. This is at least one of the reasons why mass and quantity – a one-million-strong army – is still considered important in Russia.

Whilst during the 1990s this emphasis on 'traditional' war was often perceived by Western scholars as the result of the inability of Russian strategists to move on from the Cold War, it was a reflection of the idiosyncrasies of the country's strategic priorities and threat perceptions. Especially during the 1990s and early 2000s, these differed from those of the West, where the focus had shifted to 'new wars' and counter-insurgency. This does not mean that the ability to deal with conflicts at the lower end of the conflict spectrum was not seen as important in Russia. The recognition that small wars and insurgencies around its troubled peripheries were an important source of insecurity and necessitated changes to the country's military structure dates back to the early 1990s. However, it was not until the 2008 modernization programme that this issue was addressed systematically with improvements made to rapid reaction capabilities, permanent readiness, and relevant training similar to Western developments. Although 'traditional' perceptions of potential enemies and conflicts are still a component in Russian military doctrine, it also reflects the understanding that its armed forces are required to deal with a much broader range of sources of insecurity.

The focus on the potential implications of the military revival for Russia's neighbours and the West overlooks the important domestic dimension of the phenomenon. The Kremlin's concern over threats to internal order and regime stability is an expression of fears over outside, and specifically Western, meddling in the internal affairs of sovereign states. These fears date back at least to the late 1990s. In the first decade of the twenty-first century, they were enhanced by the 'colour revolutions' in neighbouring countries. The view of outside interference as a threat to Russia's sovereignty and security has undoubtedly grown in significance in recent years. Threats to the 'information space and internal sphere' were formalized in the 2014 military doctrine as a domestic military danger for the first time. The same concern is also reflected in recent military thinking on a general level. Developments in the force structures underline the centrality of concerns over domestic and regime stability in Russian security thinking. Domestic political reasons have been a significant factor for maintaining these structures throughout the post-Soviet era. They have been used to ensure regime stability and enforce internal order, which is an important prerequisite for bolstering and upholding great power status in the eyes of the Kremlin. Force structures tasked specifically with the maintenance of public and internal order, like the FSB and the National Guard Service, continue to be strengthened. As such, the revival of Russian military power is as much about internal (in)security and stability as it is about fighting wars and the perceived need to be able to stand up against other leading powers.

Russia has also used military power, on various occasions, to cooperate and to strengthen its position in a multilateral system. Moscow's use of the military in this capacity is often overlooked. In the early 1990s, the establishment of CIS peacekeeping forces in Moldova, Georgia and Tajikistan, and cooperation with the OSCE and UN, were a part of this picture. A number of the Russian force structures, such as the Federal Security Service (FSB), the civil defence troops under the Ministry for Emergency Situations (MChS) and the now defunct Federal Service for the Control of the Drugs Trade (FSKN) have also been used for the establishment of international security cooperation. Whilst initially force structure personnel were used predominantly to make up for shortcomings in the regular armed forces in dealing with operations at the lower end of the conflict spectrum, they have since been deployed internationally to cooperate in the areas of counter-terrorism, counter-drug operations and humanitarian operations. They have cooperated with the UN, with NATO and also within the framework of the CSTO to strengthen Russia's multilateral engagement in the security field. The Kremlin's support of the global war on terrorism is another important example of multilateral security cooperation. Isolation is not in Moscow's interest and in the past, its willingness to cooperate regularly continued even at times of serious tensions. The desire for inclusion has always been an important factor in the Kremlin's foreign policy. However, its view of multilateralism as closely linked to multipolarity, where cooperation is possible only if Russia has an equal say in decisions, has often led to conflict, rather than to an improvement in relations with the West.

The prospects and limits of Russia's military revival

The idea that Russian military power now rivals the West is questionable. The 2008 modernization programme has delivered clear results in equipping the armed forces with capabilities they did not have before. As the Crimea operation showed, advances have been made most notably in the strengthening of permanent readiness and rapid reaction, which were important areas of weakness that have led to problems in the past. Improvements in command, control, joint action and coordination have also been made, as demonstrated in both Crimea and Syria, and during a number of large-scale exercises training for joint combined arms operations. A costly programme of rearmament has modernized available military equipment. Following a hiatus in the delivery and upgrading of military hardware lasting for more than a decade, the armed services' stocks of weaponry have been comprehensively replenished. The impact of technological renewal on the Russian military's ability to project power on a global level was demonstrated in Syria. At the same time, the country's capabilities for expeditionary operations remain limited. It is highly unlikely that it could supply and sustain the scale of forces required for an operation comparable to the US coalition efforts in Afghanistan and Iraq.

At the same time, it has to be borne in mind that a country's military power cannot be measured in absolute terms, but only in relation to that of others. Although the Russian military is by far the most superior player in its immediate neighbourhood, and almost always has been, it still has a long way to go to achieve the parity with the West that it desires. In terms of sheer size, the Russian military is no competitor to the world's other great powers, such as the United States and China. This disparity is exacerbated by the fact that a large part of its strength is still made up of conscripts with little training and experience. For demographic and financial reasons, the Russian military's current strength of around 800,000 cannot be substantially increased. In terms of the quality of technology available, the Russian military also continues to lag decades behind more advanced countries in areas such as precision weapons, unmanned aerial vehicles and robotics. Owing to problems with the Russian ship-building industry, its maritime power projection capabilities are also limited. The economic downturn has cast serious doubts on the long-term affordability of the rearmament programme, indicating that Russia is continuing its long history of punching above its weight when it comes to its military ambitions.

Much has been made of the success of Russian 'hybrid warfare' in Crimea. There are concerns that, even if Moscow's material military capabilities continue to lag behind the West, it has made up for these shortcomings with innovations in strategic thought. The 'newness' and prowess of 'hybrid warfare' approaches pursued in Crimea should not be overstated. Russia has a rich history of innovative military thinking, which continued to develop after the end of the Cold War. However, owing to the absence of structured military reforms throughout the 1990s, there was little evidence that this shaped military policy in a systematic manner. Russian military interventions in the CIS region and in Chechnya highlighted that, for much of the post-Soviet era, the Kremlin relied on brute force and numerical superiority, exposing the shortcomings of its capabilities compared to those of the West. The operation in Crimea showed that military modernization had led to improvements in the ability of military planners to adjust tactics to suit the specific circumstances of an operation. In Crimea, the tactics chosen suited the ends to be achieved, unlike in previous operations, when poor strategy was made up for with overwhelming firepower. However, as far as Russian military thinking goes, the conclusions to be drawn from Crimea should not be overstated. As is the case for all wars and conflicts, context is key to strategy. What works in one case cannot easily be repeated elsewhere, where the circumstances are different. Moreover, the focus on 'hybrid warfare' does not adequately reflect other developments in Russian military thinking and ambitions, as the Syria operation in the form of a conventional air campaign has demonstrated. Russian military modernization aims for full-spectrum capabilities, including competitiveness in the technological realm. The importance afforded by Russian strategists to advanced technology in modern warfare dates back to the 1970s. This idea was also at the heart of the 2008 modernization programme. The ambitions of the recent military revival can only be achieved if the expensive and increasingly unaffordable rearmament programme is delivered.

Outlook and options for Russia's neighbours and the West

In spite of the caveats offered in this book, the Kremlin's growing assertiveness and the resulting international tensions nonetheless present difficult challenges to its neighbours and to the West. With the annexation of Crimea, Russia has demonstrated its willingness to use military force for territorial gain for the first time in post-Cold War history. This has led to understandable fears about what Moscow's future intentions might be, especially amongst its closest neighbours. Russia's operations in Ukraine and in Syria have also led to tensions with the West that are incomparable in scale and scope to other crisis situations in the post-Soviet past. Russia's economic problems since 2008 have slowed down the pace of military moderni-zation, but the process is ongoing and its achievements are unlikely to be reversed.

Although this book has argued that there is little evidence of a fundamental turnaround in the ambitions driving Moscow's foreign policy, it is impossible to determine for certain what its intentions are and how the readiness to use military force will develop in the future. It is also clear that intentions can change, sometimes rapidly and in unpredictable ways. Concerns, especially by Russia's neighbouring states, are therefore justified. Moscow's preparedness to use force to uphold its dominant position in what it perceives as its sphere of influence has been a constant feature since the early 1990s. As demonstrated by the war in Georgia in 2008 and especially by the annexation of Crimea in 2014, Russia is ready to defend this position, even if it involves blatant violations of international law and leads to serious international condemnation. There is little reason to believe that Russia's vision vis-à-vis the CIS region will change in the immediate future. The continuation of the use of force in this region is therefore likely in situations where Moscow sees its interests threatened there. Likely scenarios are popular uprisings that could bring into power a regime unfriendly to Russia, or a deteriorating security situation, such as the spreading of religiously motivated extremism and terrorism in Central Asia.

As for the West, from the 1990s onwards there has been an increasing air of defiance in Russia's foreign policy. As Bobo Lo has put it: 'the feelings of inferiority that once characterized Russian elite attitudes have given way to a new militancy and, in public at least, aggressive self-confidence' (2015: xvi). According to Lo, Russian leaders recognize that their 'actions are often unpopular in the West'. However, 'for them this matters less than the restoration of national self-respect and strategic independence' (2015: 203). That said, growing assertiveness is not the same as the intention to enter into all-out confrontation. Russia's capabilities to use force beyond the CIS region remain limited in any case. Moreover, it is doubtful whether the Kremlin would risk using military force in a situation that would almost certainly lead to a war with the United States or with NATO. Such a scenario would not serve its national interests, however defined, because it would threaten the very existence of the regime and the international status Moscow has been trying so hard to protect. For the Baltic States at least this means that the likelihood of a Russian military incursion into their territories is lower than often claimed. In many ways, the idea that

Russia is seeking to confront the West militarily contradicts the most important and consistent ambition driving its foreign policy throughout history: the desire for great power recognition. Having great power status for Russia denotes international acknowledgement of its position as an important pole in the international system which, on a par with other great powers, has an equal say in important global decisions of the day. This cannot be achieved in isolation, but only in cooperation with other powerful actors.

Since 2014, fears about 'hybrid threats' emanating from Russia and directed against the West, such as 'information warfare' or 'cyber warfare' have become increasingly prominent. Such fears are not without basis, especially since evidence of attempted Russian interference in the political affairs of various Western countries – most prominently in the US presidential elections in autumn 2016 – has come to light. It is beyond doubt that these measures present a challenge to the countries concerned. Having said this, it is important to disentangle political and military threats in this respect. The tendency to couch these issues in the language of warfare unhelpfully implies that the West is already at war with Russia and that 'hybrid threats' could easily turn into military action. As such, Western discourse mirrors recent developments in Russia's own military doctrine, where outside information influence on the country's politics and society is presented as a military danger. Given the already tense international situation, Charap's suggestion that this represents a 'dangerous misuse of the word "war"' cannot be repeated often enough (2015/16: 52). Using unnecessarily militarized language to describe various tools for seeking political influence also makes it more difficult to find realistic solutions to the problem. Clearly, a military response or increases to defence budgets are unlikely to yield satisfactory results in such cases.

Realistically, there are limited options available to the West and to the international community to stop Russia's military revival outright, bar the imposition of sanctions banning the export of defence technology and dual-use equipment into Russia. The US and EU have already put such measures into place following the annexation of Crimea. It will also be difficult, as it has been in the past, to prevent Russia from using military force in certain situations in the future. Any forceful attempt to do so would considerably increase the chances of a global military conflict.

There are choices to be made, then, on deciding how to respond to an increasingly assertive Russia. When it comes to responding to uses of military force, the West can only lead by example in using war as a last resort, within the parameters of international law, and to condemn Russia in the strongest terms when it does not do the same. Moscow's military actions in Ukraine and Syria have already had serious consequences for the country's international image, even if it is clear that the country has also achieved recognition through military means. As isolation is not in Moscow's interest, and Russia continues to seek respect as a great power from the global community, there is some hope that international repercussions and condemnation will be a factor in its future decisions on when and where to use military force.

There are also choices to be made for NATO in deciding on how to alleviate the fears felt by its members that are geographically closest to Russia. These fears have been exacerbated by Moscow's military posturing and brinkmanship along its western border. Doing nothing is clearly not an option. Even if a convincing case can be made that Russian actions are not driven by 'revanchist' intentions, chance and uncertainty make such fears understandable and justified. On the one hand, NATO's emphasis on unity and resolve and the reinforcement of its position on its eastern flank make a good deal of sense, as do the efforts made by European non-NATO member states to increase military cooperation with the alliance. On the other hand, military deterrence in response to the Kremlin's actions is a double-edged sword. The major change since 2014 has been not so much the strengthening of Russian military power in relative terms. The more significant change has occurred in international views of its capabilities, which often exaggerate them. The tendency by the West to overestimate Russian capabilities has brought the Kremlin closer to its goal of great power recognition. From this point of view, the effectiveness of deterrence intended to stop military posturing is far from guaranteed. Insofar as it validates to a degree Russia's status as a powerful global actor, it could even encourage such behaviour, rather than prevent it.

The experience of the Cold War has also taught us what an ever-more intense security dilemma can lead to. As Samuel Charap and Timothy Colton noted, tensions between Russia and NATO since 2014 have already had negative repercussions for the security environment in Europe. This is because although 'the NATO moves are a response to genuine threat perceptions of East Central European allies over Russia's behaviour since 2014', Moscow nonetheless 'sees in them nothing more than a continuation of the long-running process of NATO moving its military infrastructure closer to Russia's borders. In response, Russia has announced a build-up in its Western Military District' (2017: 161–2). If the trend of uncompromising rhetoric and military posturing on both sides continues, a renewed arms race is a possible outcome. Given Russia's economic situation and comparative conventional military weakness, the West would probably win such a race yet again. However, it would nonetheless be costly for all states and societies involved and the danger of intended or unintended esca-lation in the face of spiralling tensions is worth bearing in mind.

Even if the announcement of a 'New Cold War' is premature, it is clear that the severity of tensions and depth of political differences since 2014 have made a return to 'business as usual' unacceptable not only to the West, but also to Russia. As such, Lo's assessment of the future appears not entirely unrealistic: 'Russia-West relations are set on a path of negative continuity: an overall downward trend, punctuated by periodic crises and, more rarely, brief upturns' (2015: 166). Whether this is the bottom line, or a more optimistic scenario is also possible, will depend not only on Russia. As the book has shown, although the Kremlin's foreign policy towards the West clearly has become more confrontational in recent years, its view of the West as a potential friend or foe is not all black and white. Moreover, changes in Russia's relations with the West are not predetermined, but at least in part the product of mutual interaction. As such, the difficult question NATO and the West will have to answer is whether a middle ground between a policy that might lead to another arms race with all the costs and dangers that this involves, and doing nothing, or a weak response that could be interpreted as 'appeasement' can be found. With the annexation of Crimea and Russia's ongoing involvement in, and destabilization of, Donbas in mind, diplomatic efforts to alleviate tensions and prevent spiralling confrontation are problematic for political reasons. This situation is unlikely to change in the near future, because the prospect of Russia returning Crimea to Ukraine is simply unrealistic. At the same time, it is hard to see how the negative continuity in relations can be stopped without the willingness of both sides to communicate. On a bilateral basis, cooperation in areas of common interest has resumed, as the ceasefire in Syria brokered by the US and Russia in conjunction with Jordan in July 2017 demonstrated. Although such initiatives will not solve the bigger political problems dividing the United States and Russia, and have not done so

in the past, they nonetheless are a small step towards the normaliza-
tion of relations.

For NATO as an alliance, which has to reconcile the divergent
political priorities of its members, the resumption of official cooperation with Russia has been especially problematic. Moreover, as
was the case in the past, its open-door policy and the potential for
further enlargement will always make its relationship with Russia difficult. Having said this, even when it comes to NATO, the Kremlin's
views of the alliance are not as black and white as they appear today.
Russia–West relations have 'survived' two rounds of eastward enlargement in the past, without this leading to the brink of war or even the
Kremlin's refusal of ongoing cooperation. Russia's concerns about
eastward enlargement resonated throughout the post-Cold War years.
However, in 1999, it was not the inclusion of Poland, Hungary and
the Czech Republic in NATO, but the new strategic doctrine NATO
had adopted, followed by Operation Allied Force, which led Russia to
freeze relations with the alliance. Although Russia expressed opposition to the Baltic States joining NATO, it continued its cooperation
with the alliance also in 2004. The NATO–Russia Council (NRC) had
been established in 2002 and relations had also improved based on
the common interest in fighting international terrorism. Against this
background, Putin stated in 2005, even when relations with the US
in particular had soured as a result of the 2003 war in Iraq, that 'the
choice made in favour of dialogue and co-operation with NATO was
the right one and has proved fruitful . . . In just a very short time we
have taken a gigantic step from past confrontation to working together
and from mutual accusations and stereotypes to creating modern
instruments for co-operation such as the NRC' (quoted in Forsberg
and Herd 2015: 47). Both rounds of NATO enlargement had been preceded by diplomatic efforts by the alliance to keep Russia 'on side'. As

such, this shows that future NATO enlargement need not necessarily lead to aggressive Russian action.

By 2017, all practical cooperation under the NRC remained suspended, with only a handful of meetings for 'political dialogue' having taken place since 2014. Whether pragmatic cooperation on matters of mutual interest should be resumed, as they were after the temporary suspension following the war in Georgia in 2008, is a difficult decision. On the one hand, for political reasons, official military cooperation with Russia after it annexed Crimea would be viewed as highly contentious, because it would imply that the Kremlin can get away even with flagrant violations of international law. On the other hand, it cannot be denied that Russia will continue to matter to NATO and to the West 'whether it is a partner, or, as seems more likely, a competitor', as Monaghan noted. 'New' security challenges transcending the borders of nation states, such as international terrorism and the proliferation of weapons of mass destruction, will be difficult to resolve without Russia's involvement, because of its influence and location (2016: xi). Moreover, although practical cooperation in certain areas will not lead to the solution of the bigger political differences between NATO and the West – a fact that has been obvious not only since 2014 – it ultimately might be necessary as the better alternative to escalating confrontation.

Arguably, the danger of tensions spiralling out of control is the biggest challenge to international security resulting from Russia's military revival. Although the West takes no blame for the Kremlin's aggressive behaviour or violations of international law, it bears some responsibility for how the subsequent tensions are managed. Russia will not go away, whether as a threat or as a partner, as noted above. As such, another attempt at creating the conditions required for a stable relationship, no matter how challenging, seems preferable to the acceptance that containment is the only option left. In the words of the former British ambassador to the Soviet Union, Rodric Braithwaite (2014), any attempt to alleviate tensions in East–West relations 'would involve much eating of words on all sides . . . It will be very hard to achieve. It may already be too late. But the alternatives are liable to be far worse'.

References

Adamsky, D. (2008), 'Through the looking glass: the Soviet military-technical revolution and the American revolution in military affairs', *Journal of Strategic Studies*, 31(2), 257–94.

Adamsky, D. (2010), *The Culture of Military Innovation: The Impact of Cultural Factors on the Revolution in Military Affairs in Russia, the US, and Israel*, Stanford, CA: Stanford University Press.

Adamsky, D. (2015), 'Cross-domain coercion – the current Russian art of strategy', *Proliferation Papers*, 54.

Adomeit, H. (1995), 'Russia as a "great power" in world affairs: images and reality', *International Affairs*, 71(1), 35–68.

Adomeit, H. (2009), 'Inside or outside? Russia's policies towards NATO', in E. Wilson Rowe and S. Torjesen (eds), *The Multilateral Dimension in Russian Foreign Policy*, London: Routledge.

Agence France Press (2015), 'Nordic countries extend military alliance in face of Russian aggression', *The Guardian*, 10 April, https://www.theguardian.com/world/2015/apr/10/nordic-countries-extend-military-alliance-russian-aggression

Aldis, A. and McDermott, R. (eds) (2004), *Russian Military Reform, 1992–2000*, London: Routledge.

Aleinik, L. (2006), 'MChS uidet uz armii', *Gazeta*, 21 February.

Allensworth, W. (1998), '"Derzhavnost": Aleksandr Lebed's vision for Russia', *Problems of Post-Communism*, 45(2), 51–8.

Allison, R. (1993), *Military Forces in the Soviet Successor States*, Adelphi Paper 280, London: Brassey's.

Allison, R. (2001), 'Russia and the new states of Eurasia', in A. Brown (ed.), *Contemporary Russian Politics: A Reader*, Oxford: Oxford University Press, pp. 443–54.

Allison, R. (2008), 'Russia resurgent? Moscow's campaign to "coerce Georgia to peace"', *International Affairs*, 84(6), 1145–71.

Allison, R. (2013a), *Russia, the West, and Military Intervention*, Oxford: Oxford University Press.

Allison, R. (2013b), 'Russia and Syria: explaining alignment with a regime in crisis', *International Affairs*, 89(4), 795–823.

Allison, R. (2014), 'Russian "deniable" intervention in Ukraine: how and why Russia broke the rules', *International Affairs*, 90(6), 1255–97.

Amann, M. et al. (2016), 'Putin wages hybrid war against Germany', *Der Spiegel online*, 5 February, http://www.spiegel.de/international/ europe/putin-wages-hybrid-war-on-germany-and-west-a-10754 83.html

Antonenko, O. (1999), 'Russia, NATO and European security after Kosovo', *Survival*, 41(4), 124–44.

Arbatov, A.G. (1998), 'Military reform in Russia: dilemmas, obstacles and prospects', *International Security*, 22(4), 83–134.

Arbatov, A.G. (2000), 'The transformation of Russian military doctrine. Lessons learned from Kosovo and Chechnya', The Marshall Center Papers, 2.

Aris, S. (2009), 'The Shanghai Security Organisation: "tackling the three evils"', *Europe-Asia Studies*, 61(3), 457–82.

Art, R.J. (1996), 'American foreign policy and the fungibility of force', *Security Studies*, 5(4), 7–42.

Averre, D. (2009), 'From Pristina to Tskhinvali: the legacy of Operation Allied Force in Russia's relations with the West', *International Affairs*, 85(3), 571–91.

Averre, D. and Davies, L. (2015), 'Russia, humanitarian intervention, and the responsibility to protect: the case of Syria', *International Affairs*, 91(4), 813–34.

Babaeva, S. (2004), 'V. Cherkesov: professional'naia ustanovka chekistov – ne presechenie a preduprezhdenie', *Izvestiia*, 24 March.

Babak, V. (2000), 'Russia's relations with the near abroad', *The Soviet and Post-Soviet Review*, 27(1), 93–103.

Bacon, E. (2000), 'The power ministries', in N. Robertson (ed.), *Institutions and Political Change in Russia*, Basingstoke: Macmillan.

Bacon, E. and Renz, B. (2003), 'Restructuring security in Russia: return of the KGB?', *The World Today*, 59(5), 26–7.

Bacon, E. and Renz, B. with Cooper, J. (2006), *Securitising Russia: The Domestic Politics of Putin*, Manchester: Manchester University Press.

Baev, P. (1996), *The Russian Army in a Time of Troubles*, London: Sage.

Baev, P. (1998), 'Peacekeeping and conflict management in Eurasia', in R. Allison and C. Bluth (eds), *Security Dilemmas in Russia and Eurasia*, London: The Royal Institute of International Affairs, pp. 209–29.

Baranovsky, V. (2000), 'The Kosovo factor in Russia's foreign policy', *The International Spectator*, 35(2), 113–30.

Barany, Z. (2005), 'Defence reform, Russian style: obstacles, options, opposition', *Contemporary Politics*, 11(1), 33–51.

Barany, Z. (2007), *Democratic Breakdown and Decline of the Russian Military*, Princeton, NJ: Princeton University Press.

Barkin, S.J. and Cronin, B. (1994), 'The state and the nation: changing norms and the rules of sovereignty in international relations', *International Organization*, 48(1), 107–30.

Bartles, C. (2016), 'Getting Gerasimov right', *Military Review*, January–February, 30–8.

Bartles, C. and McDermott, R. (2014), 'Russia's military operation in Crimea: road-testing rapid reaction capabilities', *Problems of Post-Communism*, 61(6), 46–63.

Baunov, A. (2015), 'The concert of Vienna: Russia's new strategy', Moscow Carnegie Centre, 8 November, http://carnegie.ru/commen tary/?fa=61892

Bellamy, A.J. (2009), *Responsibility to Protect: The Global Effort to End Mass Atrocities*, Cambridge: Polity.

Benbow, T. (2004), *The Magic Bullet? Understanding the Revolution in Military Affairs*, London: Brasseys.

Bennett, G. (2000a), 'The Ministry of Internal Affairs of the Russian Federation', Conflict Studies Research Centre Document 106.

Bennett, G. (2000b), 'The Federal Security Service of the Russian Federation', Conflict Studies Research Centre Document C102.

Bennett, C. (2000c), 'The SVR: Russia's Intelligence Service', Conflict Studies Research Centre Document 103.

Bennett, G. (2002), 'The Federal Border Guard Service', Conflict Studies Research Centre Document C107.

Bialer, S. (1981), *The Domestic Context of Soviet Foreign Policy*, Boulder, CO: Westview Press.

Biersack, J. and O'Lear, S. (2014), 'The geopolitics of Russia's annexation of Crimea: narratives, identity, silences, and energy', *Eurasian Geography and Economics*, 55(3), 247–69.

Binnendijk, H. and Herbst, J. (2015), 'Putin and the "Mariupol Test"', *The New York Times*, 15 March, https://www.nytimes.com/2015/03/16/opinion/putin-and-the-mariupol-test.html

Blair, D. (2015), 'How to protect the Baltic states?', *The Telegraph*, 19 February, http://www.telegraph.co.uk/news/worldnews/europe/russia/11423416/How-do-we-protect-the-Baltic-States.html

Blank, S. (2012), 'A work in regress? Russian defense industry and the unending crisis of the Russian state', in R. McDermott, B. Nygren and C. Vendil Pallin (eds), *The Russian Armed Forces in Transition: Economic, Geopolitical and Institutional Uncertainties*, London: Routledge, pp. 151–68.

Blank, S. (2015), 'Imperial ambitions: Russia's military build-up', *World Affairs*, May–June.

Blank, S. (2016), 'Counting down to a Russian invasion of the Baltics', *Newsweek*, 12 December, http://www.newsweek.com/counting-down-russian-invasion-baltics-414877

Blinova, E. (2005), 'MChS vstupilo v NATO. Rossiskie spasateli nedovol'ny deistviami ikh amerikanskikh kolleg na poberezh'e Meksikanskogo zaliva', *Nezavisimaia Gazeta*, 21 September.

Blua, A. (2004), 'Tajikistan: Tajiks to replace Russian border guards on Afghan border', Radio Free Europe/Radio Liberty, 4 March, https://www.rferl.org/a/1052661.html

Bluth, C. (1998), 'Russian military forces: ambitions, capabilities and constraints', in R. Allison and C. Bluth (eds), *Security Dilemmas in Russia and Eurasia*, London: Royal Institute for International Affairs.

Bodner, M. (2016), 'Changing of the guard: NATO brings in army general to deter Russia', *The Moscow Times*, 12 May, https://the moscowtimes.com/articles/changing-of-the-guard-nato-brings-in-army-general-to-deter-russia-52844

Boguslavskaya, Y. (2015), 'Russia and NATO: looking for the less pessimistic scenario', in R. Czulda and M. Madej (eds), *Newcomers No More? Contemporary NATO and the Future from the Perspective of 'Post-Cold War' Members*, Warsaw: International Relations Institute.

Borger, J. (2016), 'Finland says it is nearing security deal with US amid concerns over Russia', *The Guardian*, 22 August, https://www.theguardian.com/world/2016/aug/22/finland-us-russia-military-security

Bradshaw, M. and Connolly, R. (2016), 'Barrels and bullets: the geostrategic significance of Russia's oil and gas exports', *Bulletin of the Atomic Scientists*, 73(2), 156–64.

Braithwaite, R. (2014), 'Ukraine crisis: no wonder Vladimir Putin says Crimea is Russian', *The Independent*, 1 March, http://www.independent.co.uk/voices/comment/ukraine-crisis-no-wonder-vladimir-putin-says-crimea-is-russian-9162734.html

Breedlove, P. (2015), 'Theater Strategy', United States European Command. October, http://www.eucom.mil/media-library/document/35147/useucom-theater-strategy

Brovkin, V. (1999), 'Discourse on NATO in Russia during the Kosovo war', *Demokratizatsiya*, 7(4), 544–60.

Brown, A. (1996), *The Gorbachev Factor*, Oxford: Oxford University Press.

Brzezinski, Z. (1994), 'The premature partnership', *Foreign Affairs*, 73(2), 67–82.

Bugryi, M. (2014), 'The Crimea operation: Russian force and tactics', *Eurasia Daily Monitor*, 11(61).

Bukkvoll, T. (2009), 'Russia's military performance in Georgia', *Military Review*, 89(6), 52–69.

Bukkvoll, T. (2011), 'Iron cannot fight – the role of technology in current Russian military theory', *Journal of Strategic Studies*, 34(5), 681–706.

Bukkvoll, T. (2016a), 'Why Putin went to war: ideology, interests and decision-making in the Russian use of force in Crimea and Donbas', *Contemporary Politics*, 22(3), 267–82.

Bukkvoll, T. (2016b), 'Inefficiencies and imbalances in Russian defence spending', background paper for report, *After Hybrid Warfare – what next?* Aleksanteri Institute, University of Helsinki, http://www.helsinki.fi/aleksanteri/english/projects/VNK-report%20longerpieces.pdf

Carlsson, M., Norberg, J. and Westerlund, F. (2013), 'The military capability of Russia's armed forces in 2013', in J. Hedenskog and C. Vendil Pallin (eds), *Russian Military Capability in a Ten-Year Perspective – 2013*, Stockholm: Swedish Defence Research Agency (FOI), pp. 23–70.

Charap, S. (2013), 'Russia, Syria and the doctrine of intervention', *Survival*, 55(1), 35–41.

Charap, S. (2015/16), 'The ghost of hybrid war', *Survival*, 57(6), 51–8.

Charap, S. and Colton, T.J. (2017), *Everyone Loses: The Ukraine Crisis and the Ruinous Contest for Post-Soviet Eurasia*, London: International Institute for Strategic Studies.

Charap, S. and Darden, K. (2014), 'Russia and Ukraine', *Survival: Global Politics and Strategy*, 56(2), 7–14.

Chekinov, S.G. and Bogdanov, S.A. (2013), 'The nature and content of a new-generation war', *Military Thought*, 4, 12–23.

Cherkesov, V. (2007), Nel'zia dopustit', chtoby voiny prevratilis' v torgovtsev', *Kommersant*, 9 October.

Clunan, A. (2009), *The Social Construction of Russia's Resurgence; Aspirations, Identity and Security Interests*, New York: Johns Hopkins University Press.

Connolly, R. (2015), 'Troubled times: stagnation, sanctions and the prospects for economic reform in Russia', Chatham House Research Paper, https://www.chathamhouse.org/sites/files/chathamhouse/field/field_document/20150224TroubledTimesRussiaConnolly.pdf

Connolly, R. and Hanson, P. (2016), 'Import substitution and economic sovereignty in Russia', Chatham House Research Paper, https://www.chathamhouse.org/sites/files/chathamhouse/publications/research/2016-06-09-import-substitution-russia-connolly-hanson.pdf

Connolly, R. and Sendstad, C. (2016), 'Russian rearmament: an assessment of defence-industrial performance', *Problems of Post-Communism*, doi: 10.1080/10758216.2016.1236668

Cooper, H. and Erlanger, S. (2014), 'Military cuts render NATO less formidable deterrent to Russia', *The New York Times*, 26 March, https://www.nytimes.com/2014/03/27/world/europe/military-cuts-render-nato-less-formidable-as-deterrent-to-russia.html

Cooper, J. (1998), 'The military expenditure of the USSR and the Russian Federation 1987–97', *SIPRI Yearbook 1998*, Stockholm: Stockholm International Peace Research Institute, pp. 243–56.

Cooper, J. (2006), 'Developments in the Russian arms industry', *SIPRI Yearbook 2006*, Oxford: Oxford University Press.

Cooper, J. (2016), 'Russia's state armament programme to 2020: a quantitative assessment of implementation, 2011–2015', Swedish Defence Research Agency (FOI) Report, March.

Cordesman, A.H. (2015), 'Russia in Syria: hybrid political warfare', Centre for Strategic and International Studies Commentary, 23 September, https://www.csis.org/analysis/russia-syria-hybrid-political-warfare

Cross, S. (2002), 'Russia and NATO toward the twenty-first century: conflicts and peacekeeping in Bosnia-Herzegovina and Kosovo', *The Journal of Slavic Military Studies*, 15(2), 1–58.

Cross, S. (2006), 'Russia's relationship with the United States/NATO in the US-led global war on terrorism', *The Journal of Slavic Military Studies*, 19, 175–92.

Crow, S. (1992), 'Russian peacekeeping: defense, diplomacy, or imperialism?', RFE/RL Research Report, 1(37), 37–40.

CSIS (2016), 'Global Security Forum 2016: Welcoming remarks and plenary I – navigating 21st century security challenges', transcript, 1

December, https://www.csis.org/analysis/global-security-forum-2016-welcoming-remarks-and-plenary-i-navigating-21st-century-security

Dallinn, A. (ed.) (1960), *Soviet Conduct in World Affairs*, New York: Columbia University Press.

Davies, L. (2015), 'Russian institutional learning and regional peace operations: the cases of Georgia and Moldova', *Comilas Journal of International Relations*, 3, 81–99.

De Haas, M. (2011), 'Russia's military doctrine development (2000–10)', in S. Blank (ed.), *Russian Military Politics and Russia's 2010 Defense Doctrine*, Carlisle, PA: Strategic Studies Institute.

Dejevsky, M. (2017), 'As Syria's war enters its endgame, the risk of US-Russian conflict escalates', *The Guardian*, 21 June.

Desmond, D. (1995), 'The restructuring of the security services in post-Communist Russia, 1991–1994', *Low Intensity Conflict and Law Enforcement*, 4(1), 133–53.

Deyermond, R. (2016), 'The uses of sovereignty in twenty-first century Russian foreign policy', *Europe-Asia Studies*, 68(6), 957–84.

Dobrolyubov, N. (2013), 'Emergency at the defense ministry', *Moscow Defense Brief*, 34(2).

Donaldson, R.H. and Nogee, J.L. (2009), *The Foreign Policy of Russia: Changing Systems, Enduring Interests*, 4th edn, London: Routledge.

Echevarria, A.J. (2015), 'How we should think about "grey zone" wars', *Infinity Journal*, 5, 16–20.

Egnell, R. (2010), 'Winning hearts and minds? A critical analysis of counter-insurgency operations in Afghanistan', *Civil Wars*, 10(3), 282–303.

Facon, I. (2006), 'Integration or retrenchment? Russian approaches to peacekeeping', in R. Utley (ed.), *Major Powers and Peacekeeping: Perspectives, Priorities and the Challenges of Military Intervention*, Farnham: Ashgate, pp. 31–48.

Farmer, B. (2015), 'Putin will target the Baltic next, Defence Secretary warns', *The Telegraph*, 18 February, http://www.telegraph.co.uk/news/worldnews/vladimir-putin/11421751/Putin-will-target-the-Baltic-next-Defence-Secretary-warns.html

Federal Law 226-FZ (2016), 'O voiskakh natsional'noi gvardii Rossiiskoi Federatsii', 3 July, https://rg.ru/2016/07/06/gvardia-dok. html

Feklyunina, V. (2012), 'Image and reality: Russia's relations with the West', in R. Kanet and R. Freire (eds), *Russia and European Security*, Dordrecht: Republic of Letters Publishing, pp. 29–103.

FEMA website, 'About the agency', http://www.fema.gov/about-agency

FitzGerald, M.C. (1994), 'The Russian military's strategy for "sixth generation warfare"', *Orbis*, 38(3), 457–76.

Foreign Policy Conception of the Russian Federation 1993 (2005), in A. Melville and T. Shakleina (eds), *Russian Foreign Policy in Transition: Concepts and Realities*, Budapest: Central European University Press, pp. 27–64.

Forsberg, T. and Herd, G. (2015), 'Russia and NATO: from windows of opportunities to closed doors', *Journal of Contemporary European Studies*, 23(1), 41–57.

Forsberg, T., Heller, R. and Wolf, R. (2014), 'Status and emotions in Russian foreign policy', *Communist and Post-Communist Studies*, 47(3), 261–8.

Forster, P. (2006), 'Beslan: counter-terrorism incident command: lessons learned', *Homeland Security Affairs*, 2(3), 1–7.

FSB website (2013), '8 oktiabria 2013 goda ispolniaetsia 15 let so dnia sozdaniia Tsentra spetsial'nogo naznacheniia FSB Rossii', 8 October, http://www.fsb.ru/fsb/press/message/single.htm!id%3D10437483 @fsbMessage.html

Fuller, W.C. Jr (1992), *Strategy and Power in Russia, 1600–1914*, New York: Free Press.

Galeotti, M. (2002), 'Emergency presence', *Jane's Intelligence Review*, January.

Galeotti, M. (2013), *Russian Security and Paramilitary Forces since 1991*, Oxford: Osprey.

Galeotti, M. (2014a), 'The "Gerasimov Doctrine" and Russian non-linear war', in Moscow's Shadow blog, 6 July, https://inmoscows shadows.wordpress.com/2014/07/06/the-gerasimov-doctrine-and-russian-non-linear-war/

Galeotti, M. (2014b), 'Blowback's a bitch: MH17 and the East Ukraine campaign long-term costs for Russia', in Moscow's Shadow blog, 20 July, https://inmoscowsshadows.wordpress.com/2014/07/20/blow backs-a-bitch-mh17-and-the-east-ukraine-campaigns-long-term-costs-for-russia/

Galeotti, M. (2015), 'The West is too paranoid about Russia's "infowar"', *The Moscow Times*, 30 June, https://themoscowtimes.com/articles/the-west-is-too-paranoid-about-russias-infowar-op-ed-47802

Galeotti, M. (2016), 'Putin's new National Guard – what does it say if you need your own personal army?', Moscow's Shadows blog, 5 April, https://inmoscowsshadows.wordpress.com/2016/04/05/putins-new-national-guard-what-does-it-say-when-you-need-your-own-personal-army/

Galtung, J. (1971), 'A structural theory of imperialism', *Journal of Peace Research*, 8(2), 81–117.

Gamov, A. (2000), 'Evgenii Murov: U okhrany prezidenta net prava na oshibku', *Komsomol'skaia Pravda*, 4 August.

Gerasimov, V. (2013), 'Tsennost' nauki v predvidenii', *Voenno-promyshlennyi kur'er*, 8(476), 1–3.

Giles, K. (2016), 'Russia's "new" tools for confronting the West: continuity and innovation in Moscow's exercise of power', Chatham House Research Paper, March.

Gleason, G. (2001), 'Why Russia is in Tajikistan', *Comparative Strategy*, 20(1), 77–89.

Godzimirski, J.M. (2009), 'Russia and the OSCE: from high expectations to denial?', in E. Wilson Rowe and S. Torjesen (eds), *The Multilateral Dimension in Russian Foreign Policy*, London: Routledge.

Golts, A. (2004), 'The social and political condition of the Russian military', in S.E. Miller and D. Trenin (eds), *The Russian Military: Power and Policy*, Cambridge, MA: MIT Press, pp. 73–94.

Golts, A.M. and Putnam, T.L. (2004), 'State militarism and its legacies: why military reform has failed in Russia', *International Security*, 29(2), 121–58.

Gordon, M. (2014), 'Russia displays a new military prowess in Ukraine's East', *The New York Times*, 21 April.

Gorenburg, D. (2012), 'Challenges facing the Russian defence establishment', Russian Military Reform blog, 20 December, https://russiamil.wordpress.com/2012/12/20/challenges-facing-the-russian-defense-establishment/

Gorenburg, D. (2016), 'What Russia's military operation in Syria can tell us about advances in its capabilities', PONARS Policy Memo 424, http://www.ponarseurasia.org/memo/advances-russian-military-operations

Goure, D. (2013), 'The measure of a superpower: a two major regional contingency military for the 21st century', The Heritage Foundation Special Report, No. 128.

Gover, D. (2014), 'Probability of Russia invading Ukraine "very high" warns US intelligence', *International Business Times*, 29 March, http://www.ibtimes.co.uk/probability-russia-invading-ukraine-very-high-warns-us-intelligence-1442441

Graham, T. (2015), 'Russia's Syria surprise (and what America should do about it)', *The National Interest*, 15 September, http://nationalinterest.org/feature/russias-syria-surprise-what-america-should-do-about-it-13844

Grau, L. and Thomas, T. (1999), '"Soft Log" and concrete canyons: Russian urban combat logistics in Grozny', *Marine Corps Gazette*, 83(1), 72–4.

Gressel, G. (2015), 'Russia's quiet military revolution and what it means for Europe', European Council on Foreign Relations Policy Brief, 143, 12 October, http://www.ecfr.eu/publications/summary/russias_quiet_military_revolution_and_what_it_means_for_europe4045

Gvosdev, N.K. (2014), 'The bear awakens: Russia's military is back', *The National Interest*, 12 November, http://nationalinterest.org/commentary/russias-military-back-9181

Hagel, C. (2014a), Reagan National Defence Forum keynote speech, 15 November, https://www.defense.gov/News/Speeches/Speech-View/Article/606635/

Hagel, C. (2014b), Defence Innovation Days at the Southeastern New England Defence Industry Alliance opening keynote, 3 September, https://www.defense.gov/News/Speeches/Speech-View/Article/605602/

Hall, S. (2016), 'Why the West should pay more attention to Moldova', Stratfor Worldview, 20 February, https://www.stratfor.com/weekly/why-west-should-pay-more-attention-moldova

Harrison, M. (2008), 'Secrets, lies and half-truths: the decision to disclose Soviet defense outlays', PERSA Working Papers, 55, http://www2.warwick.ac.uk/fac/soc/economics/staff/mharrison/archive/persa/055.pdf

Headley, J. (2003), 'Sarajevo, February 1994: the first NATO-Russia crisis of the post-Cold War era', *Review of International Studies*, 29, 209–27.

Hedenskog, J., Konnander, V., Nygren, B., Oldberg, I. and Pursiainen, C. (eds) (2005), *Russia as a Great Power: Dimensions of Security under Putin*, London: Routledge.

Heistein, A. and Michlin-Shapir, V. (2016), 'Russia's hybrid warfare victory in Syria', *The National Interest*, 19 May, http://nationalinterest.org/feature/russias-hybrid-warfare-victory-syria-16273

Herd, G. (2010), 'Security strategy: sovereign democracy and Great Power aspirations', in M. Galeotti (ed.), *The Politics of Security in Modern Russia*, Farnham: Ashgate, pp. 7–28.

Herspring, D. (2005a), 'Vladimir Putin and military reform in Russia', *European Security*, 14(1), 137–55.

Herspring, D. (2005b), 'Dedovshchina in the Russian army: the problem that won't go away', *The Journal of Slavic Military Studies*, 18(4), 607–29.

Higgins, A. (2016), 'EU suspects Russian agenda in migrants' shifting arctic route', *The New York Times*, 2 April, http://www.nytimes.com/2016/04/03/world/europe/for-migrants-into-europe-a-road-less-traveled.html

Hodgson, Q. (2003), 'Is the Russian bear learning? An operational and tactical analysis of the second Chechen war, 1999–2002', *Journal of Strategic Studies*, 26(2), 64–91.

Hoffman, F. (2007), *Conflict in the 21st Century: The Rise of Hybrid Wars*, Arlington, VA: Potomac Institute.

Holmes, K. (2014), 'Putin's asymmetrical war on the West', *Foreign Policy*, 5 May, http://foreignpolicy.com/2014/05/05/putins-asym metrical-war-on-the-west/

Hopf, T. (2005), 'Identity, legitimacy, and the use of military force: Russia's Great Power identities and military intervention in Abkhazia', *Review of International Studies*, 31, 225–43.

Hosking, G. (1998), *Russia: People and Empire, 1552–1917*, Cambridge, MA: Harvard University Press.

Hosking, G. (2001), *Russia and the Russians: a History*, London: Allen Lane.

House of Commons Defence Committee (2014a), 'Oral evidence: Towards the next Defence and Security Review: Part Two: NATO', HC358, 24 June, http://data.parliament.uk/writtenevidence/com mitteeevidence . svc / evidencedocument / defence - committee / to wards-the-next-defence-and-security-review-part-two-nato/oral/ 11114.html

House of Commons Defence Committee (2014b), 'Towards the next Defence and Security Review: Part Two – NATO', Third Report of Session 2014–15, HC358, 31 July, http://www.publications.parlia ment.uk/pa/cm201415/cmselect/cmdfence/358/358.pdf

House of Lords (2015), 'The EU and Russia before and beyond the crisis in Ukraine', European Union Committee, 6th Report of Session 2014–15, HL Paper 115.

Iancu, N., Fortuna, A., Barna, C. and Teodor, M. (eds) (2015), *Countering Hybrid Threats: Lessons Learned from Ukraine*, Amsterdam: IOS Press.

Independent International Fact-Finding Mission on the Conflict in Georgia (2009), Report, September 2009, http://echr.coe.int/Docu ments/HUDOC_38263_08_Annexes_ENG.pdf

Ivanov, I. (2002), *The New Russian Diplomacy*, Washington, DC: Brookings Institution Press.

Jackson, C.F. (2016), 'Information is not a weapons system', *Journal of Strategic Studies*, 39(5–6), 820–46.

Jackson, R. (1999), 'Sovereignty in world politics: a glance at the conceptual and historical landscape', *Political Studies*, 47(3), 431–56.

Jelavich, B. (1974), *St. Petersburg and Moscow*, Bloomington, IN: Indiana University Press.

Jelavich, B. (1991), *Russia's Balkan Entanglements 1806–1914*, Cambridge: Cambridge University Press.

Johnson, D. (2015), 'Russia's approach to conflict: implications for NATO's deterrence and defence', NATO Defence College, Research paper 111.

Johnson, K. (2015), 'Putin's Mediterranean power play in Syria', *Foreign Policy*, 2 October, http://foreignpolicy.com/2015/10/02/putins-mediterranean-power-play-in-syria-navy-tartus-fleet/

Jones, M.J. and Smith, M.L.R. (2016), 'Is Vladimir Putin orchestrating Russian football hooligans to push Britain out of the EU?, *The Daily Telegraph*, 21 June, http://www.telegraph.co.uk/news/2016/06/21/is-vladimir-putin-orchestrating-russian-football-hooligans-to-pu/

Jones, R. (1984), 'Opposition to war and expansion in late eighteenth-century Russia', *Jahrbücher für Geschichte Osteuropas*, 32, 34–51.

Jones, S. (2016), 'Defence spending by NATO's Europe states up as uncertainty rises', *The Financial Times*, 30 May, https://www.ft.com/content/e0058620-259d-11e6-8ba3-cdd781d02d89

Kagan, F. and Higham R. (2002), 'Introduction', in F. Kagan and R. Higham (eds), *The Military History of Tsarist Russia*, London: Palgrave, pp. 1–10.

Kahl, C. (2017), 'The danger of war with Russia is, incredibly, more real than you'd think as US wades deeper into Syria', *The National Post*, 10 April.

Kalb, M. (2015), *Imperial Gamble: Putin, Ukraine, and the New Cold War*, Washington, DC: The Brookings Institution.

Kalyev, R. (2002), 'Can "power ministries" be reformed?, *Perspective*, 13(1), http://www.bu.edu/iscip/vol13/kaliyev.html

Katz, M. (2013), 'Russia and the conflict in Syria: four myths', *Middle East Policy*, 10(2), 38–46.

Kernen, B. and Sussex, M. (2012), 'The Russo-Georgian war: identity, intervention and norm adaptation', in M. Sussex (ed.), *Conflict in the Former Soviet Union*, Cambridge: Cambridge University Press, pp. 91–117.

Kieras, J.D. (2016), 'Irregular warfare', in D. Jordan, J.D. Kieras, D.J. Lonsdale, I. Speller, C. Tuck and C.D. Walton (eds), *Understanding Modern Warfare*, Cambridge: Cambridge University Press, pp. 299–375.

King, C. (1994), 'Eurasia letter: Moldova with a Russian face', *Foreign Policy*, 97, 106–20.

Klein, M. (2009), 'Russia's military capabilities: "Great Power" ambitions and reality', Stiftung Wissenschaft and Politik (SWP), Berlin, SWP Research Paper 12.

Klein, M. (2012), 'Towards a "New Look" of the Russian armed forces? Organizational and personnel changes', in R. McDermott, B. Nygren and C. Vendil Pallin (eds), *The Russian Armed Forces in Transition: Economic, Geopolitical and Institutional Uncertainties*, London: Routledge.

Klein, M. and Pester, K. (2014), 'Russia's armed forces on modernisation course', Stiftung Wissenschaft und Politik (SWP), Berlin, SWP Comments 9.

Knight, A. (2015), 'Why Russia needs Syria', *The New York Review of Books*, 8 October.

Kofman, M. (2016), 'Russian hybrid warfare and other dark arts', War on the Rocks blog, 11 March, http://warontherocks.com/2016/03/russian-hybrid-warfare-and-other-dark-arts/

Kofman, M. and Rojansky, M. (2015), 'A closer look at Russia's "hybrid war"', Kennan Cable, 7, https://www.wilsoncenter.org/sites/default/files/7-KENNAN%20CABLE-ROJANSKY%20KOFMAN.pdf

Kolstø, P. (2004), 'Nation building in Russia: a value-oriented strategy', in P. Kolstø and H. Blakkisrud (eds), *Nation-Building and Common Values in Russia*, Lanham, MD: Rowman and Littlefield.

Kosyrev, A. (1994), 'The lagging partnership', *Foreign Affairs*, 73(3), 59–71.

Kramer, M. (2002), 'Oversight of Russia's intelligence and security agencies: the need for and prospect for democratic control', PONARS Policy Memo No. 281.

Kramer, M. (2005), 'Guerrilla warfare, counterinsurgency and terrorism in the North Caucasus: the military dimension of the Russian-Chechen conflict', *Europe-Asia Studies*, 57(2), 209–90.

Kramer, M. (2008), 'Russian policy toward the Commonwealth of Independent States: recent trends and future prospects', *Problems of Post-Communism*, 55(6), 3–19.

Kramnik, I. and Bogdanov, K. (2016), 'Kardinal'nye gvardeitsy: v Rossii sozdali parallel'nuiu armiiu', lenta.ru news service, 6 April, https://lenta.ru/articles/2016/04/06/ng/

Kroenig, M. (2015), 'Facing reality: getting NATO ready for a new Cold War', *Survival*, 57(1), 49–70.

Kryshtanovskaya, O. and White, S. (2003), 'Putin's militocracy', *Post-Soviet Affairs*, 19(4), 289–306.

Kubicek, P. (2009), 'The Commonwealth of Independent States: an example of failed regionalism?', *Review of International Studies*, 35(1), 237–56.

Lambeth, B. (1995), 'Russia's wounded military', *Foreign Affairs*, 74(2), 86–98.

Lambeth, B. (1996), *The Warrior who would Rule Russia*, London: Brassey's.

Lavrov, A. (2015), 'Towards a professional army', *Moscow Defense Brief*, 4(48).

Lavrov, S. (2016), 'Russia's foreign policy in a historical perspective', *Russia in Global Affairs*, 30 March, http://eng.globalaffairs.ru/number/Russias-Foreign-Policy-in-a-Historical-Perspective-18067

Leander, A. (2004), 'Drafting community: understanding the fate of conscription', *Armed Forces and Society*, 30(4), 571–99.

Legvold, R. (2009), 'The role of multilateralism in Russian foreign policy', in E.W. Rowe and S. Torjesen (eds), *The Multilateral Dimension in Russian Foreign Policy*, London: Routledge.

Legvold, R. (2016), *Return to the Cold War*, Cambridge: Polity.

Lepingwell, J.W.R. (1994), 'The Russian military and security policy in the "near abroad"', *Survival*, 36(3), 70–92.

Liddell Hart, B.H. (1967), *Strategy: The Indirect Approach*, London: Faber and Faber.

Light, M. (1996), 'Foreign policy thinking', in N. Malcolm, A. Pravda, R. Allison and M. Light (eds.), *Internal Factors in Russian Foreign Policy*, Oxford: Oxford University Press, pp. 33–100.

Light, M. (2010), 'Russian foreign policy', in S. White, R. Sakwa and H.E. Hale (eds), *Developments in Russian Politics*, Basingstoke, Palgrave, pp. 225–44.

Lo, B. (2015), *Russia and the New World Disorder*, London: Chatham House.

Lukyanov, F. (2016), 'Putin's foreign policy – the quest to restore Russia's rightful place', *Russia in Global Affairs*, 4 May.

MacAskill, E. (2016), 'West and Russia on course for war, says ex-NATO deputy commander', *The Guardian*, 18 May, https://www.theguardian.com/world/2016/may/18/west-russia-on-course-for-war-nato-ex-deputy-commander

Makarychev, A. and Morozov, V. (2011), 'Multilateralism, multipolarity, and beyond: a menu of Russia's policy strategies', *Global Governance: A Review of Multilateralism and International Organizations*, 17(3), 353–73.

Mankoff, J. (2007), 'Russia and the West: taking the longer view', *Washington Quarterly*, 30(2), 123–35.

Marcus, J. (2014), 'Ukraine crisis: is Russia ready to move into eastern Ukraine?', BBC, 8 April, http://www.bbc.co.uk/news/world-europe-26940375

Markedonov, S. (2007), 'The paradoxes of Russia's Georgia policy', *Russia in Global Affairs*, 5(2), http://eng.globalaffairs.ru/number/n_8551

Masters, J. (2015), 'The Russian military', Foreign Relations Council, 28 September, https://www.cfr.org/backgrounder/russian-military

Matthews, O. (2016), 'Putin's winning in Syria – but making a powerful new enemy', *The Spectator*, 20 February, https://www.spectator.co.uk/2016/02/putins-winning-in-syria-but-making-a-powerful-new-enemy/#

McDermott, R. (2005), 'Russian border guards begin withdrawal from Tajikistan', *Eurasia Daily Monitor*, 2(78), https://jamestown.org/program/russian-border-guards-begin-withdrawal-from-tajikistan/

MChS website, 'O grazhdanskoi oborone', http://www.mchs.gov.ru/activities/Grazhdanskaja_oborona/Istorija/

MChS website, 'Spasatel'nye voinskie formirovaniia', http://www.mchs.gov.ru/ministry/other/rescue_military

Mearsheimer, J. (1994/5), 'The false promise of international institutions', *International Security*, 19(3), 5–49.

Mearsheimer, J. (2001), *The Tragedy of Great Power Politics*, New York: W.W. Norton.

Mearsheimer, J. (2010), 'Structural realism', in T. Dunne, M. Kurki and S. Smith (eds), *International Relations Theories: Discipline and Diversity*, 2nd edn, Oxford: Oxford University Press.

Medvedev, D. (2008), 'Interview given to television channels Channel One, Rossia, NTV', 31 August, http://en.kremlin.ru/events/president/transcripts/48301

Merry, E.W. (2016), 'The origins of Russia's war in Ukraine: the clash of Russian and the European "civilizational choises for Ukraine"', in E.A. Wood, W.E., Pomeranz, E.W. Merry and M. Trudolyubov (eds), *Roots of Russia's War in Ukraine*, New York: Woodrow Wilson Center Press with Columbia University Press, pp. 27–50.

Monaghan, A. (2016), *The New Politics of Russia: Interpreting Change*, Manchester: Manchester University Press.

Moran, J. (2002), *From Garrison State to Nation State: Political Power and the Russian Military under Gorbachev and Yeltsin*, Westport, CT: Praeger.

Morozov, V. (2012), 'Dmitry Medvedev's conservative modernization: reflections on the Yaroslavl speech', PONARS Eurasia Policy Memo No. 134, http://www.ponarseurasia.org/sites/default/files/policy-memos-pdf/pepm_134.pdf

Morozov, V. (2015), 'Aimed for the better, ended up with the worst: Russia and international order', *Journal on Baltic Security*, 1(1), 26–37.

Mukhin, A. (2000), *Spetssluzhby Rossii i "bol'shaia politika"*, Moscow: SPIK-Tsentr.

Nation, R.C. (1993), *Black Earth, Red Star: a History of Soviet Security Policy, 1917–1991*, Ithaca, NY: Cornell University Press.

NATO (2006), 'Backgrounder: NATO's role in civil emergency planning', http://www.igsu.ro/documente/SAEARI/NATO_CEP.pdf

NATO (2008), 'Bucharest summit declaration', press release 49, http://www.nato.int/cps/in/natohq/official_texts_8443.htm

Neumann, I. (1996), *Russia and the Idea of Europe: a Study in Identity and International Relations*, 2nd edn, London: Routledge.

Neumann, I. (2008), 'Russia as a Great Power 1815–2007', *Journal of International Relations and Development*, 11(2), 128–51.

Nichol, J. (2011), 'Russian military reform and defense policy', US Congressional Research Service Report, No. R42006, https://www.fas.org/sgp/crs/row/R42006.pdf

Nikitin, A. (2004), 'Partners in peacekeeping', NATO opinion, 1 October, http://www.nato.int/cps/en/natohq/opinions_21119.htm?selectedLocale=en

Nikitina, Y. (2012), 'The Collective Security Treaty Organization through the looking glass', *Problems of Post-Communism*, 59(3), 41–52.

Nikolsky, A. (2013a), 'The invisible reform of the Border Guard Service', *Moscow Defense Brief*, 34(2).

Nikolsky, A. (2013b), 'The Russian Federal Drug Control Service', *Moscow Defense Brief*, 38(6).

Nikolsky, A. (2013c), 'Federal Protection Service – FSO', *Moscow Defense Brief*, 36(4).

Nikolsky, A. (2014), 'Russian "spetsnaz" forces – from saboteurs to court bailiffs', *Moscow Defense Brief*, 39(1).

Nikolsky, A. (2015), 'Little, green and polite: the creation of Russian special operations forces', in C. Howard and K. Pukhov (eds), *Brothers Armed: Military Aspects of the Crisis in Ukraine*, Minneapolis, MN: East View Press, pp. 124–7.

Nikolsky, A. (2016), Russia's new National Guard: foreign, domestic and personal aspects', *Moscow Defense Brief*, 52(2).

Norberg, J. (2015), 'Training to fight: Russia's major military exercises, 2011–2014', Swedish Defence Research Agency (FOI) Report, http://foi.se/rapport?rNo=FOI-R--4128--SE

Norberg, J., Westerlund, F. and Franke, U. (2014), 'The Crimea operation: implications for future Russian military interventions', in

Granholm, N., Malminen, J., Persson, G. (eds), *A Rude Awakening: Ramifications of Russian Aggression Towards Ukraine*, Stockholm: Swedish Defence Research Agency (FOI), pp. 41–9.

Olcott, M.B. (1995), 'Sovereignty and the "near abroad"', *Orbis*, 39(3), 353–67.

Orlov, I.B. (2006), 'Derzhavnost' v russkoi politicheskoi culture: Istoriya i sovremennost', 31 August, http://www.russkie.org/index.php?module=fullitem&id=9927

Oxenstierna, S. (2016), 'Russia's defense spending and the economic decline', *Journal of Eurasian Studies*, 7, 60–70.

Page, S. (1994), 'The creation of a sphere of influence: Russia and Central Asia', *International Journal*, 49(4), 788–813.

Paoli, L. (2002), 'The price of freedom: illegal drug markets and policies in post-Soviet Russia', *Annals of the American Academy of Political and Social Science*, 582(1), 167–80.

Papkova, I. (2008), 'The freezing of historical memory? The post-Soviet Russian Orthodox Church and the Council of 1917', in M.D. Steinberg and C. Wanner (eds), *Religion, Morality, and Community in Post-Soviet Societies*, Washington, DC: Woodrow Wilson Center Press, pp. 55–84.

Perry, B. (2015), 'Non-linear warfare in Ukraine: the critical role of information operations and special operations', *Small Wars Journal*, 14 August, http://smallwarsjournal.com/print/27014

Persson, G. (2017), 'The war of the future: a conceptual framework and practical conclusions: essays on strategic thought', NATO Defense College Research Division, Russian Studies Paper No. 3/17.

Petro, N.N. and Rubinstein, A.Z. (1996), *Russian Foreign Policy: from Empire to Nation-State*, New York: Longman.

Petrov, I. (2016), 'Pomeniaiut pogony: sotrudniki OMON i SOBR poluchat status voennosluzhashchikh v 2018 godu', *Rossiiskaia gazeta*, 6980(112), 25 May, https://rg.ru/2016/05/25/omon-i-sobr-poluchat-status-voennosluzhashchih-v-2018-godu.html

Petrov, N. (2016), 'Changing of the guard: Putin's law enforcement reforms', ECPR commentary, 11 April, http://www.ecfr.eu/article/

commentary_changing_of_the_guard_putins_law_enforcement_ reforms_6084

Pifer, S. (2017), 'From order to chaos: deepening divisions in Donbas', Brookings blog, 2 May, https://www.brookings.edu/blog/ order-from-chaos/2017/05/02/deepening-division-in-donbas/

Pipes, R. (1997), 'Is Russia still an enemy?', *Foreign Affairs*, 76(5), 75–6.

Presidential Decree 976 (2004), 'Vopropsy Federal'noi sluzhby Rossiiskoi Federatsii po kontroliu za oborotom narkotikov', 28 July.

Presidential decree 157 (2016), 'Voprosy Federal'noi sluzhby voisk natsional'noi gvardii Rossiiskoi Federatsii', 5 April.

Presidential Decree 329 (2016), 'Ukaz prezidenta Rossiiskoi Federatsii o shtatnoi chislennosti Vooruzhennykh Sil Rossiiskoi Federatsii', 8 July, http://publication.pravo.gov.ru/Document/View/000120160 7080015

Pukhov, R. (2016), 'A proving ground of the future', *Russia in Global Affairs*, 2. http://eng.globalaffairs.ru/number/A-Proving- Ground-of-the-Future-18075

Putin, V. (2000), 'TV address to the citizens of Russia', 24 March, http:// en.special.kremlin.ru/events/president/transcripts/24201

Putin, V. (2006), 'Annual address to the federal assembly', 10 May, http://en.special.kremlin.ru/events/president/transcripts/23577

Putin, V. (2008), 'Press statement and answers to journalists' questions following a meeting of the Russia-NATO Council', 4 April, http:// en.kremlin.ru/events/president/transcripts/24903

Putin, V. (2012a), 'Soveshchenie poslov i postoiannykh predstavitelei Rossii', 9 July, http://kremlin.ru/events/president/news/15902

Putin, V. (2012b), 'Being strong: why Russian needs to rebuild its mili- tary', *Foreign Affairs*, 21 February.

Putin, V. (2013), 'A plea for caution from Russia', *The New York Times*, 11 September, http://www.nytimes.com/2013/09/12/opinion/putin- plea-for-caution-from-russia-on-syria.html?_r=0

Putin, V. (2014a), 'Address by the president of the Russian Federation', 18 March, http://en.kremlin.ru/events/president/news/20603

Putin, V. (2014b), 'Presidential address to the federal assembly', 4 December, http://en.kremlin.ru/events/president/news/47173

Putin, V. (2016a), 'V Rossii sozdana national'lnaia gvardia', 5 April, http://www.kremlin.ru/events/president/transcripts/51643

Putin, V. (2016b), 'Zasedanii kollegii Federal'noi sluzhby bezopasnosti', 26 February, http://kremlin.ru/events/president/transcripts/51397

Pynnöniemi, K. and Rácz, A. (eds) (2016), *Fog of Falsehood: Russian Strategy of Deception and the Conflict in Ukraine*, Helsinki: The Finnish Institute of International Affairs.

Ramani, S. (2016), 'Why Putin is escalating Russia's military build-up', *Huffington Post*, http://www.huffingtonpost.com/samuel-ramani/why-putin-is-escalating-r_b_11781280.html

Rasmussen, A.F. (2014), 'Why NATO matters to America', speech at the Brookings Institution, 19 March, http://www.nato.int/cps/en/natolive/opinions_108087.htm

Rasmussen, A F. (2016), 'Why NATO should hold no illusions about Russia's intentions', *Newsweek*, 3 July, http://www.newsweek.com/why-nato-should-hold-no-illusions-about-moscows-intentions-476258

Rathi, P. (2016), 'Sweden to boost defense spending to counter Russian threat in the Baltic', *International Business Times*, 30 December, http://www.ibtimes.com/sweden-boost-defense-spending-counter-russian-threat-baltic-2467477

Reisinger, H. and Golts, A. (2014), 'Russia's hybrid warfare: waging war below the radar of traditional collective defence', NATO Defence College, Research Paper No. 105.

Remington, T. (2009), 'Parliament and the dominant party regime', in S.K. Wegren and D.R. Herspring (eds), *After Putin's Russia: Past Imperfect, Future Uncertain*, Lanham, MD: Rowman and Littlefield, pp. 39–58.

Renz, B. (2005), 'Russia's "force structures" and the study of civil-military relations', *Journal of Slavic Military Studies*, 18(4), 559–85.

Renz, B. (2006), 'Putin's militocracy? An alternative interpretation of the role of the "siloviki" in contemporary Russian politics', *Europe-Asia Studies*, 58(6), 903–24.

Renz, B. (2007), 'Crisis response in war and peace: Russia's "Emergency

Ministry" and security sector reform', *World Defence Systems*, 17, 148–52.

Renz, B. (2010), 'Russian military reform: prospects and problems', *RUSI Journal*, 155(1), 58–62.

Renz, B. (2011), 'Traffickers, terrorists and a "new security challenge": Russian counternarcotics strategy and the Federal Service for the Control of the Drugs Trade', *Small Wars and Insurgencies*, 22(1), 55–77.

Renz, B. (2012a), 'The power ministries and security services', in G. Gill and J. Young (eds), *Routledge Handbook of Russian Politics and Society*, Abingdon: Routledge, pp. 209–19.

Renz, B. (2012b), 'Civil-military relations and Russian military modernization', in R. McDermott, B. Nygren and C. Vendil Pallin (eds), *The Russian Armed Forces in Transition: Economic, Geopolitical and Institutional Uncertainties*, London: Routledge, pp. 193–208.

Renz, B. (2014), 'Russian military capabilities after 20 years of reform', *Survival*, 56(3), 61–84.

Renz, B. (2016a), 'Why Russia is reviving its conventional military power', *Parameters*, 46(2), 23–37.

Renz, B. (2016b), 'Russia and "hybrid warfare"', *Contemporary Politics*, 22(3), 283–300.

Renz, B. and Smith, H. (2016), 'Russia and hybrid warfare: going beyond the label', Aleksanteri Papers, University of Helsinki: Kikimora Publications.

Renz, B. and Thornton, R. (2012), 'Russian military modernization: cause, course and consequences', *Problems of Post-Communism*, 49(1), 44–54.

Rich, P.B. (2009), 'Introduction: the global significance of a small war', *Small Wars and Insurgencies*, 20(2), 239–50.

Rieber, A.J. (1993), 'Persistent factors in Russian foreign policy: an interpretive essay', in H. Ragsdale (ed.), *Imperial Russian Foreign Policy*, New York: Cambridge University Press, pp. 315–59.

Rieber, A.J. (2007), 'How persistent are persistent factors?', in R. Legvold (ed.), *Russian Foreign Policy in the 21st Century and the Shadow of the Past*, New York: Columbia University Press, pp. 205–78.

Roffey, R. (2016), 'Russia's EMERCOM: Managing emergencies and political credibility', Swedish Defence Research Agency (FOI) Report, May, www.foi.se/ReportFiles/foir_4269.pdf

Rotar, I. (1992), 'Two days in the life of Andrei Kozyrev', *Nezavisimaya Gazeta*, 7 November, p. 3.

Rudolph, S. (2005), 'Sovereignty and territorial borders in a global age', *International Studies Review*, 7(1), 1–20.

Rumer, E.B. and Wallander, C.A. (2003), 'Russia: power in weakness?', *The Washington Quarterly*, 27(1), 57–73.

Rushev, P. (2006), 'Zasada na svoikh', *Gazeta*, 4 June.

RUSI (2007), 'Russia's post-Beslan counterterrorism reforms', RUSI Monitor, 19 November, https://rusi.org/publication/russias-post-beslan-counterterrorism-reforms

Rutland, P. (2008), 'Russia as an energy superpower', *New Political Economy*, 13(2), 203–10.

Rutland, P. (2014), 'A paradigm shift in Russia's foreign policy', *The Moscow Times*, 18 May, https://themoscowtimes.com/articles/a-paradigm-shift-in-russias-foreign-policy-35534

Saradzhyan, S. (2006), 'Russia's system to combat terrorism and its application in Chechnya', in R. Orttung and A. Makarychev (eds), *National Counterterrorism Strategies: Legal, Institutional and Public Policy Dimensions in the US, UK, France, Turkey and Russia*, Amsterdam: IOS Press, pp. 176–90.

Saunders, P.J. (2015), 'Why America can't stop Russia's hybrid warfare', *The National Interest*, 23 June, http://nationalinterest.org/feature/why-america-cant-stop-russias-hybrid-warfare-13166

Schaffer Conroy, M. (1990), 'Abuse of drugs other than alcohol and tobacco in the Soviet Union', *Soviet Studies*, 42(3), 447–80.

Scheipers, S. (2016), 'Winning wars without battles: hybrid warfare and other indirect approaches in the history of strategic thought', in B. Renz and H. Smith, *Russia and Hybrid Warfare – Going Beyond the Label*, Helsinki: Aleksanteri Papers, 1, pp. 48–52.

Sergunin, A. (2007), *International Relations in Post-Soviet Russia: Trends and Problems*, Nizhnu Novgorod: Nizhny Novgorod Linguistic University Press.

Seton-Watson, H. (1989), *The Russian Empire, 1801–1917*, Oxford: Oxford University Press.

Sharkov, D. (2016), 'Russia is biggest threat to global security says Poland', *Newsweek*, 17 June, http://europe.newsweek.com/russia-biggest-threat-global-security-says-poland-471620?rm=eu

Shear, M. and Baker, P. (2014), 'Obama renewing US commitment to NATO alliance', *The New York Times*, 26 March, https://www.nytimes.com/2014/03/27/world/europe/obama-europe.html

Shearman, P. and Sussex, M. (2009), 'The roots of Russian conduct', *Small Wars and Insurgencies*, 20(2), 251–75.

Shlykov, V. (2004), 'The economics of defense in Russia and the legacy of structural militarization', in S.E. Miller and D. Trenin (eds), *The Russian Military: Power and Policy*, Cambridge, MA: MIT Press, pp. 157–83.

Shoigu, S. (2015), 'Meeting with Defence Minister Shoigu', Kremlin website, 7 October, http://en.kremlin.ru/events/president/tran scripts/50458

Sinovets, P. and Renz, B. (2015), 'Russia's 2014 military doctrine and beyond: threat perceptions, capabilities and ambitions', NATO Defense College Research Division, Research Paper No. 117, July.

SIPRI Military Expenditure Database 1988–2015. https://www.sipri. org/databases/milex

Smith, H. (2014), *Russian Greatpowerness: Foreign Policy, the Two Chechen Wars and International Organisations*, Helsinki: Helsinki University.

Smith, H. (2016), 'Putin's third term and Russia as a Great Power', in M. Suslov and M. Bassin (eds), *Eurasia 2.0: Russian Geopolitics in the Age of New Media*, London: Rowman and Littlefield.

Smith, J. (2016), 'Old habits, new realities: Russia and Central Asia from the end of the Soviet Union up to 9/11', in H. Rytövuori-Apunen (ed.), *The Regional Security Puzzle around Afghanistan. Bordering Practices in Central Asia and Beyond*, Leverkusen: Barbara Budrich Publishers.

Snegovaya, M. (2015), 'Putin's information warfare in Ukraine: Soviet origins of Russia's hybrid warfare', Institute for the Study of War, Russia Report No. 1.

Snyder, J. (1987/8), 'The Gorbachev revolution: a waning of Soviet expansionism?', *International Security*, 12(3), 93–131.

Snyder, J. (1991), *Myths of Empire – Domestic Politics and International Ambition*, Ithaca, NY: Cornell University Press.

Snyder, J. (1994), 'Russian backwardness and the future of Europe', *Daedalus*, 123(2), 179–201.

Soldatov, A. and Borogan, I. (2011), *The New Nobility: The Restoration of Russia's Security State and the Enduring Legacy of the KGB*, New York: Public Affairs.

Spivak, A. and Pridemore W. (2004), 'Conscription and reform in the Russian army', *Problems of Post-Communism*, 51(6), 33–43.

Spruds, A. et al. (2016), 'Internet trolling as a hybrid warfare tool: the case of Latvia', Report, NATO Stratcom Centre for Excellence, http://www.stratcomcoe.org/internet-trolling-hybrid-warfare-tool-case-latvia-0

Stent, A. (2009), 'Restoration and revolution in Putin's foreign policy', *Europe-Asia Studies*, 60(6), 1089–106.

Stent, A. (2016), 'Putin's power play in Syria: how to respond to Russia's intervention', *Foreign Affairs*, 95, 106–13.

Stent, A. and Shevtsova, L. (2002), 'America, Russia and Europe: a realignment?', *Survival*, 44(4), 121–34.

Stepanova, E. (2005), 'The use of Russia's security structures in the post-conflict environment', in A. Schnabel and G. Erhart (eds), *Security Sector Reform and Post-Conflict Peacebuilding*, New York: United Nations University Press, pp. 133–55.

Strachan, H. (2013), *The Direction of War: Contemporary Strategy in Historical Perspective*, Cambridge: Cambridge University Press.

Suny, R.G. (2001), 'The Empire strikes out: imperial Russia, "national" identity, and theories of empire', in R.G. Suny and T. Martin (eds), *A State of Nations: Empire and Nation-Making in the Age of Lenin and Stalin*, Oxford: Oxford University Press.

Suny, R.G. (2007), 'Living in the hood: Russia, empire, and old and new neighbors', in R. Legvold (ed.), *Russian Foreign Policy in the 21st*

Century and the Shadow of the Past, New York: Columbia University Press.

Suny, R.G. (2012), 'The contradictions of identity: being Soviet and national in the USSR and after', in M. Bassin and C. Kelly (eds), *Soviet and Post-Soviet Identities,* Cambridge: Cambridge University Press, pp. 17–36.

Taheri, A. (2015), 'Putin is turning the Syrian coast into another Crimea', *New York Post,* 19 September, http://nypost.com/2015/09/19/putin-is-turning-the-syrian-coast-into-another-crimea/

Taia Global (2015), 'Russian Federal Security Service (FSB) internet operations against Ukraine', Taia Global Report, https://taia.global/wp-content/uploads/2015/04/FSB-IO-UKRAINE.pdf

Taylor, B. (2003), *Politics and the Russian Army: Civil-Military Relations, 1689–2000,* Cambridge: Cambridge University Press.

Taylor, B. (2007), *Russia's Power Ministries: Coercion and Commerce,* Institute for National Security and Counterterrorism, Syracuse University.

Taylor, B. (2011), *State Building in Putin's Russia: Policing and Coercion after Communism,* Cambridge: Cambridge University Press.

The Basic Provisions of the Military Doctrine of the Russian Federation (1993), 2 November. Unofficial translation available at http://fas.org/nuke/guide/russia/doctrine/russia-mil-doc.html

The Military Balance (2016), 'Chapter ten: country comparisons – commitments, force levels and economics', *The Military Balance,* 116(1), 481–92.

The Military Doctrine of the Russian Federation (2000), 21 April. Unofficial translation available at https://www.armscontrol.org/act/2000_05/dc3ma00

The Military Doctrine of the Russian Federation (2010), 5 February. Unofficial translation available at http://carnegieendowment.org/files/2010russia_military_doctrine.pdf

The Military Doctrine of the Russian Federation (2014), 25 December. Official translation available at http://rusemb.org.uk/press/2029

The Moscow Times (2016), 'Putin's personal army: analysts on Russia's

National Guard', 7 April, https://themoscowtimes.com/articles/putins-personal-army-analysts-on-russias-national-guard-52445

Thomas, N. and Tow, W.T. (2002), 'The utility of human security: sovereignty and humanitarian intervention', *Security Dialogue*, 33(2), 177–92.

Thomas, T. (1995), 'Emercom: Russia's emergency response team', *Low Intensity Conflict and Law Enforcement*, 4(2), 227–36.

Thomas, T. (1996), 'Russian views on information-based warfare', *Airpower Journal*, Special edition, pp. 25–36.

Thomas, T. (2016), 'The evolution of Russian military thought: integrating hybrid, new-generation and new-type thinking', *Journal of Slavic Military Studies*, 29(4), 554–75.

Thornton, R. (2013), 'There is no one left to draft: the strategic and political consequences of Russian attempts to end conscription', *Journal of Slavic Military Studies*, 26(2), 219–41.

Thornton, R. (2015), 'The changing nature of modern warfare', *The RUSI Journal*, 160(4), 40–8.

Tiessalo, R. (2015), 'Russian border anxiety grows as Finland beefs up military budget', 2 December, https://www.bloomberg.com/news/articles/2015-12-02/russian-border-anxiety-grows-as-finland-beefs-up-military-budget.

Tolz, V. (2001), *Russia: Inventing the Nation*. London: Bloomsbury Press.

Torjesen, S. (2009), 'Russia as a military great power: the uses of the CSTO and the SCO in Central Asia', in E. Wilson Rowe and S. Torjesen (eds), *The Multilateral Dimension in Russian Foreign Policy*, London: Routledge.

Trenin, D. (2001), 'Desiatiletie nevyuchennykh urokov: grazhdansko-voennye otnosheniia v 90-e gody i perspektivy novogo raunda voennykh reform', *Politiia*, 20(2), 72–82.

Trenin, D. (2011), *Post-Imperium: A Eurasian Story*, Washington, DC: Carnegie Endowment for International Peace.

Trenin, D. (2016a), 'A five-year outlook for Russian foreign policy: demands, drivers and influences'. Moscow Carnegie Center, 18 March, http://carnegieendowment.org/2016/03/18/five-year-out

look-for-russian-foreign-policy-demands-drivers-and-influences-
pub-63075

Trenin, D. (2016b), *Should We Fear Russia?* Cambridge: Polity.

Trenin, D. (2016c), 'The revival of the Russian military: how Moscow
reloaded', *Foreign Affairs*, 95(3), 23–9.

Trenin, D. and Malashenko, A.V. (2004), *Russia's Restless Frontier: The
Chechnya Factor in Post-Soviet Russia*, Washington, DC: Carnegie
Endowment for International Peace.

Tsygankov, A. (1997), 'From international institutionalism to revolutionary expansionism: the foreign policy discourse of contemporary
Russia', *Mershon International Studies Review*, 41(2), 247–68.

Tsygankov, A. (2008), 'Russia's international assertiveness: what does
it mean for the West?', *Problems of Post-Communism*, 55(2), 38–55.

Tsygankov, A. (2009), 'Russia in global governance: multipolarity
or multilateralism?', in D. Lesage and P. Vercauteren (eds), *Contemporary Global Governance: Multipolarity vs New Discourses on
Global Governance*, Brussels: P.I.E. Peter Lang.

Tsygankov, A. (2016), *Russia's Foreign Policy: Change and Continuity
in National Identity*, 4th edn, Lanham, MD: Rowman and Littlefield.

Tsygankov, A. and Tarver-Wahlquist, M. (2009), 'Duelling honors:
power, identity and the Russia–Georgia divide', *Foreign Policy
Analysis*, 5, 307–26.

Tsymbal, V. and Zatsepin, V. (2015), 'A new Russian national
defence control system: reform or imitation?', *Russian Economic
Developments*, 5, 45–7.

Ulam, A.B. (1968), *Expansion and Coexistence: the History of Soviet
Foreign Policy, 1917–1967*, New York: Praeger.

United Nations (2008), 'Illicit drug trends in the Russian Federation',
UNODC Regional Office for Russia and Belarus', http://www.unodc.
org/documents/regional/central-asia/Illicit%20Drug%20Trends%
20Report_Russia.pdf

Van Herpen, M. (2015), *Putin's Wars: The Rise of Russia's New
Imperialism*, London: Rowman & Littlefield.

Vendil Pallin, C. (2007), 'The Russian power ministries: tools and insur-
ance of power', *Journal of Slavic Military Studies*, 20(1), 1–25.

Vendil Pallin, C. (2009), *Russian Military Reform: A Failed Exercise in Defence Decision Making*, London: Routledge.

Vendil Pallin, C. (2015), 'Russia challenges the West in Ukraine', *Journal on Baltic Security*, 1(1), 14–25.

Vendil Pallin, C. and Westerlund, F. (2009), 'Russia's war in Georgia: lessons and consequences', *Small Wars and Insurgencies*, 20(2), 400–24.

Weitz, R. (2014), 'NATO must adapt to counter Russia's next generation warfare', *World Politics Review*, 5 August, http://www.worldpoliticsreview.com/articles/13976/nato-must-adapt-to-counter-russia-s-next-generation-warfare

Westerlund, F. (2012), 'The defence industry', in C. Vendil Pallin (ed.), *Russian Military Capability in a Ten-Year Perspective – 2011*, Stockholm: Swedish Defence Research Agency (FOI), pp. 65–96.

Westerlund, F. and Norberg, J. (2016), 'Military means for non-military measures: the Russian approach to the use of armed force', *Journal of Slavic Military Studies*, 29(4), 576–601.

Wilson Rowe, E. and Torjesen, S. (2009), 'Key features of Russian multilateralism', in E. Wilson Rowe and S. Torjesen (eds), *The Multilateral Dimension in Russian Foreign Policy*, London: Routledge, pp. 1–20.

Wydra, D. (2004), 'The Crimea conundrum: the tug of war between Russia and Ukraine on the questions of autonomy and self-determination', *International Journal on Minority and Group Rights*, 10, 111–30.

Yeltsin, B. (1999), 'Changes in the international climate', *Vital Speeches of the Day*, 66(5), 132–3.

Yermolaev, M. (2000), 'Russia's international peacekeeping and conflict management in the post-Soviet environment', in M. Malan (ed.), *Boundaries of Peace Support Operations*, Pretoria: Institute for Security Studies.

Zatsepin, V. (2012), 'The economics of Russian defence policy: in search for the roots of inefficiency', in R. McDermott, B. Nygren and C. Vendil Pallin (eds), *The Russian Armed Forces in Transition:*

Economic, Geopolitical and Institutional Uncertainties, London: Routledge.

Ziegler, C.E. (2012), 'Conceptualizing sovereignty in Russian foreign policy: realist and constructivist perspectives', *International Politics*, 49(4), 400–17.

Index

oil and gas prices 54, 74, 75
Okhrana company 103
OMON forces 103, 104, 105
Operation Allied Force 33, 35, 43, 60, 138–42, 204
Organisation for Security and Cooperation in Europe (OSCE) 46
Orlov, Igor 23
Oxenstierna, Susanne 76

PAK-FA fighter 78
Pakistan 45
Panetta, Leon 6, 121
Partnership for Peace 95
peacekeeping operations
  CIS 39, 44, 124–34, 136, 143, 157, 162, 191–2, 197, 199
  force structures and 95, 96
  United Nations 41
Peter the Great 27
Petrov, Nikolay 115
police service 91, 92
Polish–Soviet War 29
power ministries *see* force structures
Presidential Directorate for Administrative Affairs 108
Presidential Guards Regiment 107
Presidential Protection Service (SBP) 107, 116, 117
prison service 108
procurement plans 74–5, 77–81
  legacy systems 81
  State Armament Programme 2011–25 74, 77–8, 80, 81, 82
psychological operations 178, 182
Pukhov, Ruslan 3, 80
Putin, Vladimir
  Assad, support of 154, 155
  on the collapse of the Soviet Union 121
  on cooperation with NATO 204
  Crimea speech 151
  director of FSB 89, 99
  and the force structures 90, 115, 119
  foreign policy rhetoric 146
  on humanitarian interventions 35

KGB career background 89
military reform policies 26, 32, 51, 61–2, 65, 84, 139
and multilateralism 41
on NATO enlargement 146
seeks to restore great power status 6, 24, 26, 33, 61–2, 121, 139
on sovereignty 31, 32, 35
State of the Nation speeches 31

quasi-military organizations 58, 88
see also force structures

rapid reaction forces 54, 63, 71–2, 103, 193
Rasmussen, Anders Fogh 7, 9
rearmament programme 78, 84, 198
Red Army victories 29
regime change, externally driven 35, 154
regime stability *see* domestic order and regime stability
research and development (R&D) 60, 77
revanchism 8, 122, 123, 125, 156–7, 201
revolution (1905) 28
revolution (1917) 37
Revolution in Military Affairs (RMA) 60, 164, 176–7, 179
Rice, Condoleezza 145
Rieber, Alfred J. 20, 21
Rumer, Eugene 2
Russian Empire 36–7, 40–1, 48, 61
Russian Federation, emergence of 22, 50
Russian Rescue Corps 92–3, 117
Russo–Japanese War (1904–5) 28
Russo–Turkish War (1877–8) 28
Rutland, Peter 27
Rutskoi, Aleksandr 127–8

Saakashvili, Mikheil 143, 145
Saudi Arabia 73
Saunders, Paul J. 185
Scaparrotti, General Curtis 10-11
Scheipers, Sibylle 184
Second World War 29, 41